ACTIVITY MANUAL
FOR
INTRODUCTION TO KINESIOLOGY

THE SCIENCE OF HUMAN PHYSICAL ACTIVITY

Second Revised First Edition

By Marilyn Mitchell, David Anderson,
Cassandra Stewart, and Jenny O

Bassim Hamadeh, CEO and Publisher
Michael Simpson, Vice President of Acquisitions
Jamie Giganti, Managing Editor
Jess Busch, Senior Graphic Designer
Luiz Ferreira, Licensing Specialist

Copyright © 2014 by Cognella, Inc. All rights reserved. No part of this publication may be reprinted, reproduced, transmitted, or utilized in any form or by any electronic, mechanical, or other means, now known or hereafter invented, including photocopying, microfilming, and recording, or in any information retrieval system without the written permission of Cognella, Inc.

First published in the United States of America in 2014 by Cognella, Inc.

Trademark Notice: Product or corporate names may be trademarks or registered trademarks, and are used only for identification and explanation without intent to infringe.

[Cover Image]: Copyright © by Michael Slobodian. Reprinted with permission.

Printed in the United States of America

ISBN: 978-1-62661-449-9 (pbk) / 978-1-62661-450-5 (br) / 978-1-63487-197-6 (pf)

www.cognella.com 800.200.3908

CONTENTS

ACKNOWLEDGMENTS   v

PREFACE   vii

CHAPTER ONE
Kinesiology: The Emergence of a New Field of Study   1
- *Activity #1.1: Career Options*   1
- *Activity #1.2: Measuring Performance*   7
- *Lab #1: Employing a Cross-Disciplinary Approach to Examining Physical Activity*   11

CHAPTER TWO
The History of Kinesiology   15
- *Activity #2: Important People in the Development of Kinesiology*   15
- *Lab #2: How Much Do People Know About Kinesiology?*   19

CHAPTER THREE
Anatomical and Physiological Systems   25
- *Activity #3: Measuring Heart Rate*   25
- *Lab #3.1: Analysis of Movement*   31
- *Lab #3.2: Flexibility and Range of Motion*   35

CHAPTER FOUR
Exercise Physiology Foundations   41
- *Activity #4: Caloric Intake and Expenditure*   41
- *Lab #4: The Cooper Test*   53

CHAPTER FIVE
Biomechanical Foundations   **61**
Activity #5: Biomechanical Analysis of Tools   *61*
Lab #5.1: Length-Tension Relationship in Muscles   *65*
Lab #5.2: The Physics of Stability   *71*

CHAPTER SIX
Motor Control and Motor Learning Foundations   **77**
Activity #6: Instructions and Observation in Learning a Complex Motor Skill   *77*
Lab #6.1: Hick-Hyman Law: Simple vs Choice Reaction Time   *87*
Lab #6.2: Augmented Feedback: Performance and Learning   *95*

CHAPTER SEVEN
Psychological Foundations   **103**
Activity #7: Motivation and Physical Activity   *103*
Lab #7.1: Goal Setting, Feedback, and Performance   *109*
Lab #7.2: Observational Learning and Mental Rehearsal   *115*

CHAPTER EIGHT
Developmental Foundations   **121**
Activity #8: Comparing Male and Female World Records   *121*
Lab #8.1: Assessing Fundamental Motor Skills   *127*
Lab #8.2: Task Analysis, Classification, and Skill   *133*

CHAPTER NINE
Sociocultural Foundations   **141**
Activity #9: The Walkability of Your Neighborhood   *141*
Lab #9: Gender and Race Ideologies in Sports Media   *149*

CHAPTER TEN
Epilogue   **155**
Activity #10: Exploring the American Kinesiology Association   *155*
Lab #10: Designing an Integrative Research Project   *159*

Acknowledgments

Putting together this activity manual required the contributions from several individuals. Foremost were the contributions from my co-authors, Dr. David Anderson, Cassandra Stewart and Dr. Jenny O, all having extensive teaching experience in a wide variety of courses in Kinesiology. Without their efforts, this manual would have been much more difficult to develop. In addition, we would also like to thank Dr. Nicole Bolter, Dr. Mi-Sook Kim and Dr. Maria Veri, faculty members in the Department of Kinesiology at San Francisco State University for their suggestions on the lab activities. We would also like to express our gratitude to the team at University Readers and Cognella Academic Publishing for their guidance and support in putting this manual together. Melissa Accornero was the first person at University Readers to contact me about the possibility of publishing my textbook, Introduction to Kinesiology: The Science of Human Physical Activity. I would like to thank her for encouraging me to publish this manual to accompany the textbook. There are others at University Readers who have greatly assisted in the production of the manual including Kevin Fahey, Jessica Knott, Sharon Hermann, Henry Fuentes, and Stephen Milano. Many thanks to Michael Slobodian who provided the photo of my daughter, Makaila, for the cover of the manual.

Preface

Teaching an introductory class in kinesiology is challenging. Part of the challenge is understanding the broad scope of the field, from anatomy and physiology of the human body to the sociocultural factors affecting the participation in physical activity. As in any introductory class, keeping the material current is also challenging but exciting as new and interesting discoveries in all the sub-fields of kinesiology are occurring frequently. Another challenge is providing students practical experiences to enhance the lecture material. The notion that we learn best by doing should not be lost on the Kinesiology Student. At the University of Colorado at Boulder, where I first developed an Introduction to Kinesiology course, classes were typically anywhere from 300 to 400 students in the lecture. Unfortunately we did not have resources for separate labs or tutorials. While the course was popular, I always felt that ideally, separate lab sections would greatly enhance the student experience. Fortunately, in our department of Kinesiology at San Francisco State University, we have had separate laboratory sections for many years. In the process, we have developed lab experiences in many of the sub-fields in kinesiology. Several of the lab experiences in the manual were taken from those that we have successfully conducted over the last several years. In developing this manual, my co-authors and I provide laboratory experiences in all the major sub-fields in kinesiology. We have recognized, however, that many departments of kinesiology do not have lab sections. To help overcome this limitation, we also include in this manual practical activities designed for the individual student. For both the labs and the activities, we provide practical experiences that do not rely upon expensive equipment. As most instructors know, labs can be set up in different ways. We might suggest that instructors can enhance students' learning by making the students responsible for conducting the labs. For example, a group of students can introduce the background of a lab, collect and analyze the data for that lab, and then give a presentation of the findings to the class using the discussion questions in the lab as a basis for highlighting the relevance and importance of the material. For those instructors without separate lab sections, the activities for the individual students that are provided can also result in the students not only turning in their assignments to the instructor but also reporting their results in the lecture class or posting their results on the class's online platform (Blackboard, ILearn, etc.). Finally, some of the labs and activities in this manual promote an understanding of the cross-disciplinary nature of kinesiology, an approach that is emphasized in the textbook. For example, the lab experience in Chapter 7, "Goal-Setting, Feedback, and Performance," requires an understanding of concepts in both sport psychology and motor learning. We have also tried to make the labs and activities a little more stimulating by providing figures, graphs, and tables where necessary. Finally, both the labs and activities conclude with discussion questions that are designed to stimulate critical thinking and encourage further research into the topic at hand. On behalf of my co-authors and me, we hope that the manual will serve as a relevant and helpful addition to the lecture material.

Chapter One
Kinesiology: The Emergence of a New Field of Study

CHAPTER ACTIVITY #1.1

CAREER OPTIONS

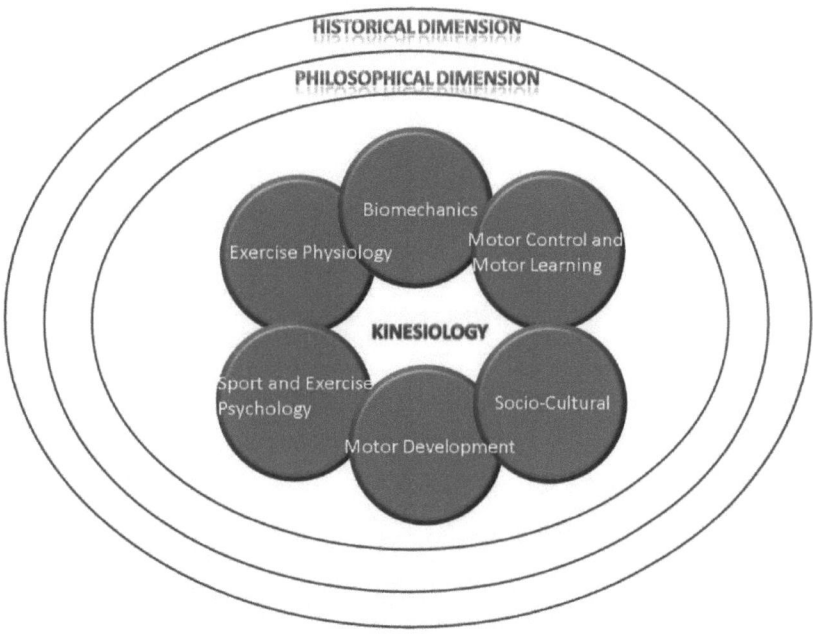

PURPOSE

The purpose of this activity is to review material presented in Chapter 1 of the textbook (Mitchell, 2013) and to investigate career options in kinesiology.

Name_____
Section_____ Date_____

INTRODUCTION

As discussed in the textbook, kinesiology is the study of human movement and is a rather large field encompassing many subfields. It is important for the kinesiology student to gain an appreciation of all of the subfields and how they are interrelated. In addition, with an undergraduate degree in kinesiology, the student can begin to consider several career opportunities. This activity allows the student to review many of the important aspects of kinesiology as well as providing the student an opportunity to investigate a career option of their interest.

Name_____
Section_____ Date_____

CHAPTER SUMMARY QUESTIONS

1. In addition to the historical and philosophical dimensions of kinesiology, what are the major subfields within kinesiology?

   1. _____

   2. _____

   3. _____

   4. _____

   5. _____

   6. _____

2. Provide three examples of each type of movement (sportive, symbolic, and supportive) described by Charles (1994):

   Sportive

   1. _____

   2. _____

   3. _____

   Symbolic

   1. _____

   2. _____

   3. _____

   Supportive

   1. _____

   2. _____

   3. _____

Name_____

Section_____ Date_____

3. Provide two examples of declarative and procedural knowledge in kinesiology:

Declarative

1. _____

2. _____

Procedural

1. _____

2. _____

4. List the six levels of analyses for kinesiology:

1. _____

2. _____

3. _____

4. _____

5. _____

6. _____

5. In the space provided below, in your own words, defend why no one level of analysis is more important than another in understanding how a movement skill is performed.

Name_____
Section_____ Date_____

CHAPTER ACTIVITY

Equipment and Materials Needed: Access to a Library and the Internet

On Pages 13–18 in the textbook, there are a number of possible career options listed and described in kinesiology. Pick one career option and answer the following questions. You may have to use the textbook or other references such as articles, other textbooks, the Internet, etc. One excellent reference is <u>Careers in Sports, Fitness and Exercise</u> sponsored by the America Kinesiology Association (2011).

_____
Career Option

References Used (provide the link for Internet references)

_____

_____

_____

For this career, what environment would you be working in?

What would be your major responsibilities, roles, and functions?

Whom would you work with?

What types of personal skills and abilities would you need to succeed?

What education and certifications would you need?

What is the outlook of this career for the near and long-term future?

Name_____

Section_____ Date_____

REFERENCES

American Kinesiology Association (2011). *Careers in sport, fitness, and exercise: An authoritative guide to land the job of your dreams.* Champaign, IL: Human Kinetics.

Charles, J. (1994). *Contemporary Kinesiology: An introduction to the study of human movement in higher education.* Englewood, CO: Morton.

Mitchell, M. (2013). *Introduction to kinesiology: The science of human physical activity.* San Diego, CA: Cognella Academic Publishing.

Name_____
Section_____ Date_____

CHAPTER ACTIVITY #1.2

MEASURING PERFORMANCE

Source: http://commons.wikimedia.org/wiki/File:Darts_in_a_dartboard.jpg. Copyright in the Public Domain.

PURPOSE

The aim of this activity is to introduce methods for reducing data and simplifying their interpretation. The focus is on measuring and analyzing the outcomes of a series of attempts at a task. In virtually all subfields of kinesiology (Mitchell, 2013), measurements are taken to help describe the activities of the individual performance or behavior. We often try to summarize performances by using measurements that describe the average and the variability of the individual or within a group of individuals doing the same task. This activity allows the student to become more familiar with three very common measurements of human performance: absolute, constant, and variable error.

INTRODUCTION

Studying human behavior usually involves the collection and assessment of large amounts of data. Performance outcomes are some of the most common measures taken in sports and physical activity. Examples of outcomes include the distance a ball is thrown or a performer jumps, the time it takes to complete a task, or the error relative to a particular goal such as hitting a target with an arrow. When data are collected for multiple trials, the first step in data reduction often involves blocking consecutive trials into groups and determining a dependent measure that best represents the groups of trials. For example, think of the number of different ways you could block the scores obtained from throwing 20 consecutive balls at a target (e.g., 1 block of 20, 2 blocks of 10, etc.). The size of the block is determined by the purpose of the data collection and the need to keep the data manageable without obscuring important trends.

Once trials are blocked, measures of central tendency and variability can be determined. Common measures of central tendency include the *mean, median,* and *mode*. These measures reflect the common tendencies for the data. The mean (or average) is calculated by adding all the scores and dividing by the number of scores. The median is found by ranking the data in ascending or descending order and locating the "middle" score. The median is often used when there are "outliers" in the data (e.g., some scores may be unusually large or small, perhaps because the performer was not paying attention on those trials). The median is often a more representative measure in these cases because it is not sensitive to outliers. Finally, the mode is the most frequently occurring score.

Variability refers to how scores are spread relative to each other. Common measures include the *range, standard deviation,* and *coefficient of variation*. The range is simply the difference between the highest and lowest scores. The standard deviation is a measure of how far away, on average, the scores are from the mean. A formula to calculate the

Name_____
Section_____ Date_____

standard deviation is given later. The coefficient of variation is calculated by dividing the standard deviation by the mean.

Because human behavior is goal-directed, performance is often evaluated by the magnitude of error associated with each trial (i.e., how far away was each trial from the goal). We usually describe errors in terms of their magnitude (how far away) and direction (e.g., did we undershoot or overshoot the target, move too fast or too slow). Often a plus sign (+) is used to indicate that we moved too far or too slowly relative to the goal, and a minus sign (–) is used to indicate that we didn't move far enough or moved too quickly.

When trials are blocked into groups, three common error measures can be determined: *absolute error* (AE), *constant error* (CE), and *variable error* (VE) (Henry, 1974; Newell, 1976; Schutz and Roy, 1973; Spray, 1986). AE is a measure of the overall magnitude of error and is calculated by summing (Σ) the absolute value of the error scores $|e|$ for each trial and dividing by the number of trials (n). The formula for AE is given below.

$$AE = \frac{\sum_{k=1}^{n} |e|}{n}$$

CE is a measure of bias in relation to the goal and is calculated by summing the error scores for each trial and dividing by the number of trials. The formula for CE (below) is exactly the same as that for AE, only the positive and negative signs are taken into account. Note that the formula for calculating CE is the same as that for the mean.

$$CE = \frac{\sum_{k=1}^{n} (e)}{n}$$

VE is a measure of variability in the data and it represents how, on average, the scores are spread around the mean. It is calculated by adding up the difference between each score and the mean (CE in this case) for each trial and dividing by the number of trials. Note that because some of the individual differences will be positive and some negative, we square each difference before adding them, then divide by n, and ultimately take the square root of that number. The formula for VE (which is the same as that for the standard deviation, except that we divide by n instead of n – 1) is given below.

$$VE = \sqrt{\frac{\sum_{k=1}^{n} (e - CE)^2}{n}}$$

MATERIALS/METHOD

Participants

Everyone in the class can participate in this activity.

Name_____
Section_____ Date_____

Task and Apparatus

The task is to calculate AE, VE, and CE for the three participants' data in the table below. The data represent the scores for five trials of a throwing task, where negative scores indicate target undershooting and positive scores indicate target overshooting.

Procedure

The calculations can be done by hand, with a calculator, or by inputting the data into a spreadsheet.

Data Analysis

Calculate AE, CE, and VE for each participant and insert the values into the appropriate cells of the table.

TRIALS	Participant 1	Participant 2	Participant 3
Trial 1	-5	5	1
Trial 2	5	5	1
Trial 3	4	4	-1
Trial 4	-4	4	0
Trial 5	0	0	1
$\Sigma \|e\|$			
$\Sigma (e)$			
AE			
CE			
$\Sigma (e - CE)^2$			
VE			

DISCUSSION QUESTIONS

1. Which participant had the largest magnitude of error and which participant had the smallest magnitude of error?

Name_____
Section_____ Date_____

2. Which participant(s) tended to overshoot the target and which participant(s) tended to undershoot the target?

3. Which participant was the most consistent and which participant was the least consistent?

4. Why might we want to use more than one type of error to characterize performance?

5. What are some of the limitations of using absolute error?

6. How would you characterize accuracy, bias, and consistency with two-dimensional error scores (e.g., when a throw could undershoot the target or overshoot the target and land to the left or the right of the target)?

REFERENCES

Hancock, G. R., Butler, M. S., and Fischman, M. G. (1995). On the problem of two-dimensional error scores: Measures and analyses of accuracy, bias, and consistency. *Journal of Motor Behavior, 27,* 241–250.

Henry, F. M. (1974). Variable and constant performance errors within a group of individuals. *Journal of Motor Behavior, 6,* 149–154.

Mitchell, M. (2013). *Introduction to kinesiology: The science of human physical activity.* San Diego, CA: Cognella Academic Publishing.

Newell, K. M. (1976). More on absolute error, etc. *Journal of Motor Behavior, 8,* 139–142.

Schutz, R. W. and Roy, E. A. (1973). Absolute error: The devil in disguise. *Journal of Motor Behavior, 5,* 141–153.

Spray, J. A. (1986). Absolute error revisited: An accuracy indicator in disguise. *Journal of Motor Behavior, 18,* 225–238.

Name_____
Section_____ Date_____

CHAPTER LAB #1

EMPLOYING A CROSS-DISCIPLINARY APPROACH TO EXAMINING PHYSICAL ACTIVITY

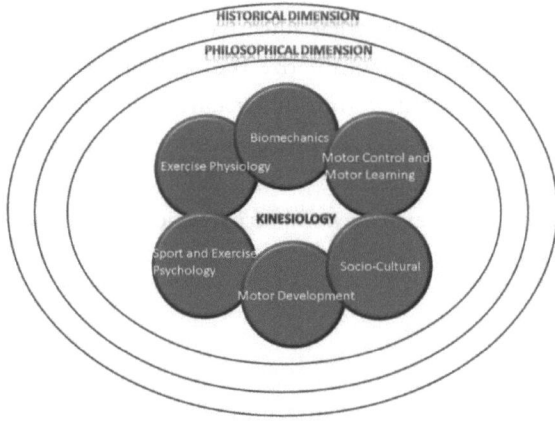

PURPOSE

The aim of this lab is to examine a kinesiology-related topic using a cross-disciplinary approach.

INTRODUCTION

In Chapter 1 of the textbook (Mitchell, 2013) it was suggested that using a cross-disciplinary approach to studying thematic problems in kinesiology may lead to a more comprehensive understanding of how and why we engage in physical activity. This means that instead of taking a relatively narrow approach to examining a kinesiology-related issue (i.e., examining it within only *one* subfield of kinesiology), we can choose to incorporate several subfields into our examination. Certainly, each of the subfields in kinesiology can be seen as highly specialized and able to provide us with an abundance of information about a specific issue. There is a great deal of merit, however, in pursuing cross-disciplinary research, given the assumption that the various subfields are known to interact with, and thus, influence one another.

MATERIALS/METHOD

Participants

Students will work in groups of two or three to complete this lab assignment. Every member of the class should participate.

Task and Apparatus

The activity requires students to examine a specific kinesiology-related research question using a cross-disciplinary approach.

Name_____
Section_____ Date_____

Procedure

On Page 9 of the text, Mitchell (2013) provides the following example of a thematic problem in kinesiology: "How does one improve cardiovascular fitness?" Students must select <u>three</u> subfields in kinesiology (i.e., biomechanics, motor control and motor learning, sociocultural, motor development, sport and exercise psychology, and/or exercise physiology) and explore this problem using Internet and library search methods.

Students are expected to find <u>one</u> peer-reviewed journal article from <u>each</u> subfield that specifically addresses cardiovascular fitness in some way. Each journal article must have a publication date of 2005 or more recent (i.e., through the current year). Students must then summarize the findings (i.e., the results) of each journal article, and clearly explain how these finding contribute to our understanding of factors that may influence cardiovascular fitness (use the table provided in the results section of this lab outline). Students must also complete the discussion questions at the end of this lab outline and hand them in with their results section table.

*Regarding use of the Internet to find peer-reviewed journal articles, we recommend using your college library website's journal article search engine.

Data Analysis

Be sure to read each journal article you find. In particular, for each article it is important that you can identify what research question was asked (the purpose of the article), who the participants were, what the task was, how it was measured, and what the major findings were (only report the findings related to cardiovascular fitness). You should also be able to summarize what each article teaches you about how one improves cardiovascular fitness.

RESULTS

Identify which subfield each article is from →	Article 1 Subfield:	Article 2 Subfield:	Article 3 Subfield:
Research Question			
Participants			
Task			
Measures			
Major findings			
How findings inform our understanding of how one improves cardiovascular function			

Name_____
Section_____ Date_____

DISCUSSION QUESTIONS

1. Based on the information you gleaned through this activity, why is it important to sometimes take a broader perspective when trying to learn about a particular topic or issue?

2. Justify your choice of the each of the subfields in which you chose to examine cardiovascular fitness.

3. With respect to writing style, focus, statistical procedures or detail, and/or discussion section content, identify a few (at least two) differences you found between the three articles you read.

4. As it relates to what we understand about cardiovascular fitness, what do you think is a potential limitation of only reading one article (versus many) from each subfield?

Note: Keep your journal articles and your Chapter 1 lab work! This information will be used again in the Chapter 10 lab.

REFERENCE

Mitchell, M. (2013). *Introduction to kinesiology: The science of human physical activity.* San Diego, CA: Cognella Academic Publishing.

Name_____
Section_____ Date_____

Chapter Two
The History of Kinesiology

CHAPTER ACTIVITY #2:
IMPORTANT PEOPLE IN THE DEVELOPMENT OF KINESIOLOGY

Copyright © MatthiasKabel (CC BY-SA 3.0) at http://commons.wikimedia.org/wiki/File:Greek_vase_with_different_sportsmen.jpg.

PURPOSE

The purpose of this activity is to review material presented in Chapter 2 of the textbook (Mitchell, 2013) and to investigate some of the important people who have influenced the development of the field of Kinesiology.

INTRODUCTION

The history of Kinesiology is a fascinating account of the important events and people that helped shape the development of the field. Some of these events occurred early in human history and many occurred later in several early human civilizations. As pointed out in the text (Mitchell, 2013), Kinesiology as a field of study developed out of the field of Physical Education, as the focus of interest in many university departments shifted from teacher training and coaching to the scientific study of human movement. In this activity, we explore several events and important people that led to the development of Kinesiology as a field of study.

Name_____

Section_____ Date_____

CHAPTER SUMMARY QUESTIONS

1. Name at least two ancient civilizations that promoted dance as a form of physical activity.

    - _____

    - _____

2. Describe two differences between the athletic contests of ancient Greece and Rome.

    - _____

    - _____

3. Name three sporting events today that are similar to the ancient Roman athletic events called 'spectacular.'

    - _____

    - _____

    - _____

4. Describe three characteristics of the 'New Physical Education' in the early 20th century that distinguished it from the gymnastic systems brought to the United States from Europe in the 19th century.

    - _____

    - _____

    - _____

5. Describe two commonalities between Franklin Henry's view of physical education and Jerry Barham's view of kinesiology.

    - _____

    - _____

Name_____
Section_____ Date_____

CHAPTER ACTIVITY

Pick two important individuals who either contributed to scientific inquiry in the study of the human body or mind, or contributed to the development of Physical Education or Kinesiology as a field of study and answer the following questions. The first individual should be from a time period before 1800 and the second individual should have lived sometime after 1800. Please provide all references that you used to support your answers (e.g., our textbook, other textbooks or articles; provide the links if using Internet websites).

References Used (provide the link for Internet references)

_____

_____

_____

1. _____
   First Individual's Name

2. _____
   Date of Birth/Death

3. What was their contribution?

1. _____
   Second Individual's Name

2. _____
   Date of Birth/Death

3. What was their contribution?

REFERENCE

Mitchell, M. (2013). *Introduction to kinesiology: The science of human physical activity*. San Diego, CA: Cognella Academic Publishing.

Name_____

Section_____ Date_____

Name_____
Section_____ Date_____

CHAPTER LAB #2: HOW MUCH DO PEOPLE KNOW ABOUT KINESIOLOGY?

PURPOSE
The aim of this lab is to examine how much a more generalized (i.e., non-kinesiology) population knows about the discipline of kinesiology.

INTRODUCTION
In Chapter 2 of the textbook (Mitchell, 2013), we chronicled the development of kinesiology throughout history. You learned that the study of human movement is not a new phenomenon; in fact, it dates back to ancient times. Despite this long history, however, kinesiology is still a relatively obscure field. Being a student in a kinesiology course, you are learning much information that may be unknown to many of your peers, your family, and to others. It would be interesting to determine how much those who have not taken a kinesiology course actually know about kinesiology.

MATERIALS/METHOD

Participants
Students will work in groups of two or three to complete this lab assignment. Every member of the class should participate.

Task and Apparatus
Students will be given class time to go out on campus and "poll" students regarding kinesiology-based questions based on materials covered in the course text in Chapters 1 and 2. Students will need to have their lab outline as well as paper and pencil with them while polling their peers.

Procedure
Working in pairs, go out and about your college campus. Ask five non-kinesiology students, faculty, or staff members the questions listed below.

Try very hard not to help people with the answers—we are interested in learning how much people know about kinesiology. It's okay if someone answers, "I don't know." Record that as their answer.

Each person you question must not have already been approached by any of your classmates. Be sure to introduce yourself, be polite, and explain that this is a class assignment for your Introduction to Kinesiology course.

Name_____

Section_____ Date_____

Name_____
Section_____ Date_____

Questions (ask five people *each* question; you can tell them the answers *after* you get their responses):

1. **In a sentence, what is kinesiology?**

    Answer: *The scientific study of human movement.*

2. **What careers are available to people with a kinesiology degree?**

    Answer: *Any career related to human movement. Examples include medical doctor, physiotherapist, athletic therapist, physical education teacher, coach, exercise specialist, nutritionist, sport management specialist, etc.*

3. **When and where did the Olympics originate?**

    Answer: *776 BC in ancient Greece (first recorded Olympic Games). Involved all city-states in Greece (not a worldwide competition). Poor and wealthy competed.*

4. **When did the "fitness boom" hit?**

    Answer: *In the 1970s—leisure time physical activity such as going to the gym, jogging, aerobics, etc. did not really exist before then.*

5. **Approximately what percentage of Americans is classified as obese (BMI > 30)?**

    Answer: *34% as of 2008 (statistically accounting for measurement error anywhere from 31 to 36%).*

Data Analysis

Write your responses in the space provided below and on the back of this page to your course instructor before you leave for the day. Please print legibly.

_____

_____

_____

_____

_____

_____

_____

_____

_____

_____

Name_____

Section_____ Date_____

Name_____
Section_____ Date_____

RESULTS
Your instructor will prepare a summary of your study "findings."

Discussion Questions

1. Overall, how much did non-kinesiology majors know about the field of kinesiology?

2. Did they know about some areas more so than others?

3. Would your results have been different if you had interviewed kinesiology majors?

4. How many kinesiology majors are on your campus?

5. Which kinesiology-related careers are most popular and which careers do you think will become more/less popular in the future? Why?

6. How might the field of kinesiology better promote itself?

7. How popular is the study of physical activity and participation in physical activity today, compared with other periods in history?

REFERENCE

Mitchell, M. (2013). *Introduction to kinesiology: The science of human physical activity.* San Diego, CA: Cognella Academic Publishing.

Name_____

Section_____ Date_____

Chapter Three
Anatomical and Physiological Systems

CHAPTER ACTIVITY #3

MEASURING HEART RATE

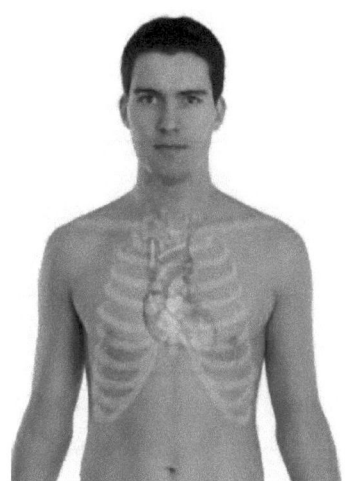

Source: http://upload.wikimedia.org/wikipedia/commons/archive/6/6b/20120628202718!Surface_anatomy_of_the_heart.png. Copyright in the Public Domain.

PURPOSE
The purpose of this activity is to review material presented in Chapter 3 of the textbook (Mitchell, 2013) and to better understand heart rate, an importance indicator of the heart's activity, and factors that can affect it.

INTRODUCTION
There are many physiological systems of the human body (Wingerd, 2007). Your textbook (Mitchell, 2013) focuses on four that play a major role in the body's ability to produce movement: the skeletal, muscular, nervous, cardiovascular and respiratory systems. In this activity we review some of the major concepts in Chapter 3, and also focus on an important measure of the cardiovascular system—the heart rate. Heart rate is defined as the number of beats (contractions) of the heart per minute. At rest, for the average adult, heart rate is approximately 70 beats per minute. But there are many factors that affect heart rate. After answering the follow questions from Chapter 3, you will do an activity that allows you to explore factors affecting heart rate.

Name_____

Section_____ Date_____

Name_____
Section_____ Date_____

CHAPTER SUMMARY QUESTIONS

1. Using the scientific terminology of joint movement, describe the action of the following joints:

- The elbow joint as you reach forward to grasp a coffee cup

   _____

- The knee joint as you stand up from sitting in a chair

   _____

- The shoulder joint as you draw a large circle on a chalkboard

   _____

2. From the anatomical position, identify which axis of rotation and plane of motion are involved in the following movements:

   Elbow flexion and extension

   _____   _____
   axis of rotation   plane of motion

   Wrist abduction

   _____   _____
   axis of rotation   plane of motion

   A ballerina performing a pirouette (spinning motion of the entire body while en pointe (tip toe)

   _____   _____
   axis of rotation   plane of motion

3. In performing a bench press (weight-lifting exercise), identify which muscles in the arms are used to push the weight up.

   _____

   Describe which part of the movement involves concentric contraction of this muscle group, and which part uses eccentric contraction.

   Concentric: _____

   Eccentric: _____

Anatomical and Physiological Systems

Name_____
Section_____ Date_____

4. If you received a "pinched nerve" as a result of any injury and you lost sensation in a particular part of the body, what part of the spinal nerve was likely affected?

_____

Briefly explain your answer:

_____

_____

5. Provide one practical example of a motor task where the size principle of motor unit recruitment would be used:

_____

_____

6. Why do you think it is more difficult to breathe at high altitudes? (Hint: You must take into account the partial pressure of oxygen in the atmosphere as well as in the lungs.)

_____

_____

Name_____
Section_____ Date_____

CHAPTER ACTIVITY

Equipment and Materials Needed: Access to the Internet

Go to the following website and read the material on heart rate: **http://en.wikipedia.org/wiki/Heart_rate**

Name three places on the body for measuring heart rate:

_____

_____

_____

What is the following quantity supposed to measure (220 – age)?

Is this formula accurate and based upon solid experimental data?

Briefly explain your answer.

Measure your resting heart rate for three consecutive days, preferably in the morning, right after you wake up. What was your average heart rate for the three days? How much physical activity do you normally do, and do you think your average heart rate reflects your physical fitness level? Why or why not?

_____

_____

_____

_____

_____

REFERENCES

Mitchell, M. (2013). *Introduction to kinesiology: The science of human physical activity.* San Diego, CA: Cognella Academic Publishing.

http://en.wikipedia.org/wiki/Heart_rate

Name_____

Section_____ Date_____

Name_____
Section_____ Date_____

CHAPTER LAB #3.1 ANALYSIS OF MOVEMENT

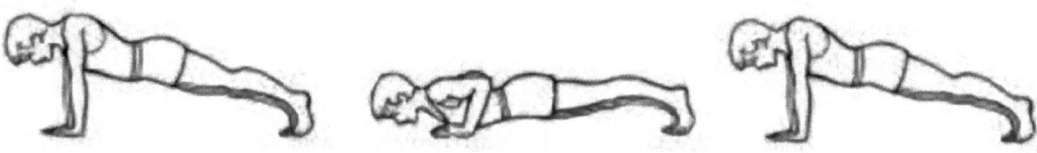

Source: http://en.wikipedia.org/wiki/File:Push_up_%28PSF%29.png. Copyright in the Public Domain.

PURPOSE

To introduce how to anatomically analyze a specific but simple human movement. You will understand the planes of motion and axes of rotation in movement as well as use critical events and phases as a start for analyzing the motion. In addition, you will understand skeletal joint movements and muscular contractions in the motion.

INTRODUCTION

An anatomical analysis of a movement breaks a skill into smaller parts. First, the motion needs to be broken into events and phases. Phases are subsections of motion within the entire movement. Some phases are obvious based on the motion. For example, a baseball pitch has a windup phase, a throwing phase where the arm moves forward, and a follow-through phase after ball release. Some phases are not as obvious in other skills, but some type of division must be made to make the analysis manageable. Where a phase appropriately starts and ends is critical for every motion. These starting and end points are the events within the motion (Hamilton, Weimar, and Luttgens, 2012). This means that every phase is sandwiched between two events and a movement can have many phases. In the case of the baseball pitch, there were three phases, which means there are four events.

After the events and phases have been identified, the joints involved, and the joint actions, the planes and axes need to be identified as well (Hamilton et al., 2012).

Name_____
Section_____ Date_____

Planes of motion

- Frontal plane (anteroposterior or AP plane): bisects the body laterally from side to side, dividing it into front and back halve (abduction, adduction).
- Sagittal plane (lateral plane): bisects the body from front to back, dividing into right and left symmetrical halves (flexion, extension, dorsiflexion, plantarflexion, hyper flexion/extension).
- Transverse plane: divides the body horizontally into superior and inferior halves (pronation, supination, rotation, horizontal abd/add).

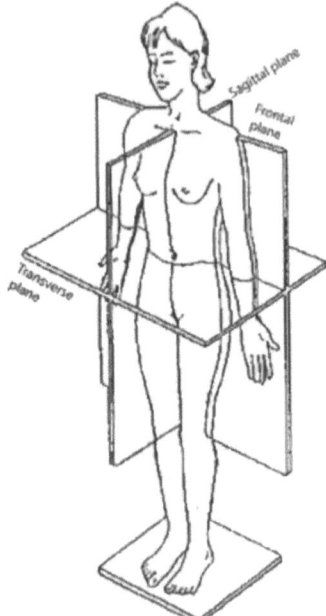

> There are three traditional planes of motion that correspond to three dimensional space, and each plane is perpendicular to each other.

Axes of rotation

- Anteroposterior: Perpendicular with the frontal plane. Passes through the body from front to back.
- Transverse: Perpendicular with the sagittal plane. Passes through the body from side to side.
- Longitudinal axis: Perpendicular with the transverse plane. Passes through the body from top to bottom.

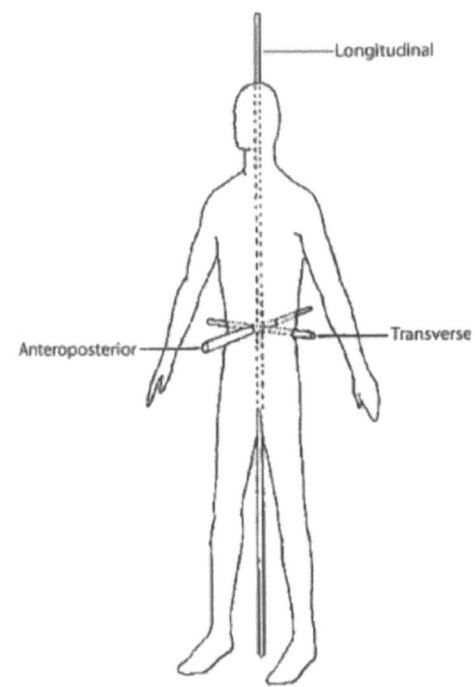

> As movement occurs in a given plane, the joint moves or turns about an axis that has a 90-degree relationship to that plane.

Name_____
Section_____ Date_____

Movements in planes

These motions are in reference to the anatomical standing position. In this position, the individual is erect, with the elbows fully extended and palms facing forward. The feet are hip width apart with the toes pointing forward (Hamilton et al., 2012).

- flexion/extension (sagittal)
- hyperflexion/hyperextension (sagittal)
- dorsiflexion/plantarflexion (sagittal)
- abduction/adduction (frontal)
- rotation (inward/outward, medial/lateral, internal/external) (transverse)
- supination/pronation (transverse)
- horizontal abduction/adduction (transverse)

MATERIAL/METHODS

Participants

The participants in this study will be your classmates.

Task and Apparatus

Procedure

In groups of five or six, perform an anatomical analysis on a simple skill (e.g., leg press, lat pull-down, jump, or push-up). Identify the events and phases, joints involved, joint actions, and the plane and axis each motion is occurring in. Make an organized chart containing this information.

DATA ANALYSIS

Example of your results section table:

Table 1. *Squat Motion Analysis*

Event	Phase	Joint	Joint Action	Plane/Axis
Event 1 Anatomical Position				
	Phase 1 Descending	Hip Knee Ankle	Flexion Flexion Dorsiflexion	All sagittal plane/ Transverse axis
Event 2 Lowest point				
	Phase 2 Ascending	Hip Knee Ankle	Extension Extension Plantarflexion	All sagittal plane/ Transverse axis
Event 3 **Return to anatomical position**				

Name_____
Section_____ Date_____

DISCUSSION QUESTIONS

1. Knowledge of kinesiology has a threefold concern for every practitioner. What are these three concerns?

2. Explain the three concerns and give specific examples of their uses by kinesiology practitioners.

3. In the example analysis of the squat, add in the muscles that are producing this motion. Explain what the contraction type is that is occurring in these muscles.

4. Every motor skill has a primary purpose and can be classified. What is the primary purpose for a squat and how would you classify it? Justify your answer.

5. Determine whether a squat is "simultaneous or sequential" in nature of the motion, and then identify its "underlying mechanics objective." Justify your answers.

REFERENCES

Hamilton, N., Weimar, W., and Luttgens, K. (2012). *Kinesiology: Scientific basis of human movement* (12th ed.). New York: McGraw-Hill.

Mitchell, M. (2013). *Introduction to kinesiology: The science of human physical activity*. San Diego, CA: Cognella Academic Publishing.

Name_____
Section_____ Date_____

CHAPTER LAB #3.2 FLEXIBILITY AND RANGE OF MOTION

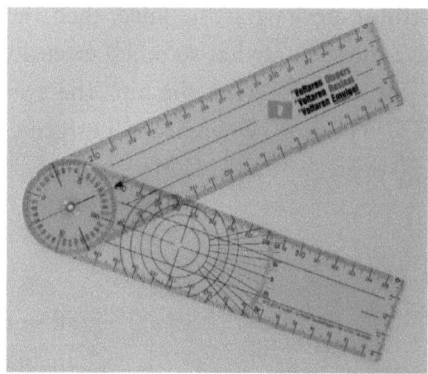

Copyright © 2010 by Depositphotos/Nebojsa78. Copyright © Voxymoron (CC BY-SA 3.0) at http://commons.wikimedia.org/wiki/File:Medizinischer_Goniometer.jpg.

PURPOSE

The purpose of this experiment is to investigate range of motion (ROM) and flexibility of a muscle. This will be done by looking at the flexibility of the hamstrings in active and passive movement and how it changes the ROM at the hip joint.

INTRODUCTION

The term "flexibility" is commonly used to describe the range of motion of a joint or a series of joints and is influenced by ligaments, tendons, bones, and muscles (Anderson and Burke, 1991). A component to flexibility is extensibility. Muscles have an extensible quality, meaning that they can be stretched beyond their normal resting length, similar to a rubber band (Mitchell, 2013, p. 79). At any point in time, a muscle has a certain amount of extensibility and flexibility available.

ROM at a joint can either be of two conditions: active movement, or passive movement. Active ROM is the amount of motion that occurs at a joint where the participant voluntarily moves the body part through the motion without assistance. Passive ROM is the amount of motion that occurs due to another person or another external force while the participant relaxes (Clarkson, 2000, pp. 4–9).

The position of the knee also has an effect on the ROM at the hip joint. This is due to the hamstring muscles being biarticular. A biarticular muscle is a muscle that crosses and acts on more than one joint (Floyd, 2012). The hamstring muscles run down the back of the thigh. There are three hamstring muscles: the semitendinous, semimembranosus, and biceps femoris.

All three muscles start at the bottom of the pelvis at a place called the ischial tuberosity. They cross the knee joint and end at the lower leg. Hamstring muscle fibers join with the tough, connective tissue of the hamstring tendons

Name_____
Section_____ Date_____

near the points where the tendons attach to bones. The hamstring muscles cross posteriorly at the hip and knee, therefore the position of the knee will affect how much ROM will be available at the hip (Floyd, 2012).

If the hamstrings are long at the knee, then there is only so much more stretching they can do at the hip. Remember that a muscle only has so much extensibility at any time. Therefore, shortening the hamstrings at the knee will allow for more ROM at the hip. The average ROM for hip flexion actively with the knee extended is 100–125 degrees (Hamilton, Weimar, and Luttgens, 2012). The average ROM for active hip flexion with the knee flexed is 115–150 degrees (Roach, 1991).

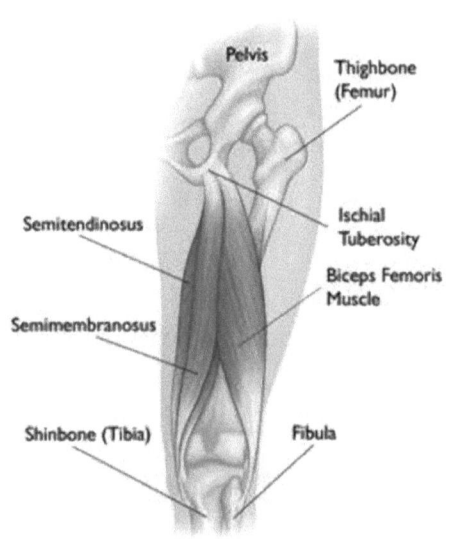

Normal hamstring anatomy

Source: R.T. Floyd, Manual of Structural Kinesiology. Copyright © 2012 by McGraw Hill.

As shown in this figure, there are several hamstring muscles that affect the ROM and flexibility of the hip joint. Because these muscles are biarticular (i.e., cross and act at more than one joint, namely the hip and knee), the position of the knee joint will affect the ROM and flexibility of the hip joint.

MATERIALS/METHODS

Participants

The participants in this study will be your classmates.

Task and Apparatus

- Goniometer
- Loose-fitting clothes

Procedure

Break into pairs where one person is the subject and the other person takes the measurements. Have the subject lie supine in the anatomical position. Place the fulcrum of the goniometer on the greater trochanter of the femur. The stationary arm is positioned along the lateral midline of the abdomen, using the pelvis for reference, and the moving arm along the lateral midline of the femur.

Name_____
Section_____ Date_____

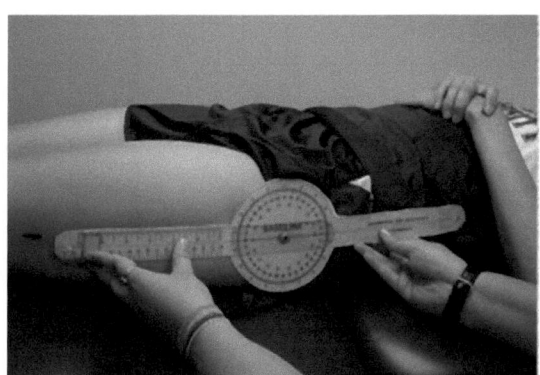 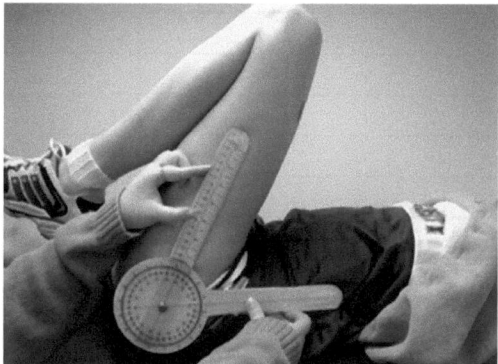

On the left panel, the ROM is measured with the knee extended: this is an example of passive motion. On the right panel, the ROM is measured with the knee flexed: an example of active motion. While the ROM varies from person to person, it is expected that ROM will be greater when the knee is flexed. With the knee flexed the average ROM for active hip flexion is 115-150 degrees. The subject shown above displays an active ROM of approximately 80 degrees.

Have the subject go through the required range of motion. Flex only the dominate leg and move as far as possible. Measure the range of motion and document it. There are four conditions that each participant will perform:

1. AKE: the subject in supine position with right knee extended active
2. PKE: the subject in supine position with right knee extended passive
3. AKF: the subject in supine position with right knee flexed active
4. PKF: the subject in supine position with right knee flexed passive

DATA ANALYSIS

Participant #	AKE	PKE	AKF	PKF
Participant #	AKE	PKE	AKF	PKF

Anatomical and Physiological Systems

Name_____

Section_____ Date_____

Averages				

Plot the class averages (example)

Name_____
Section_____ Date_____

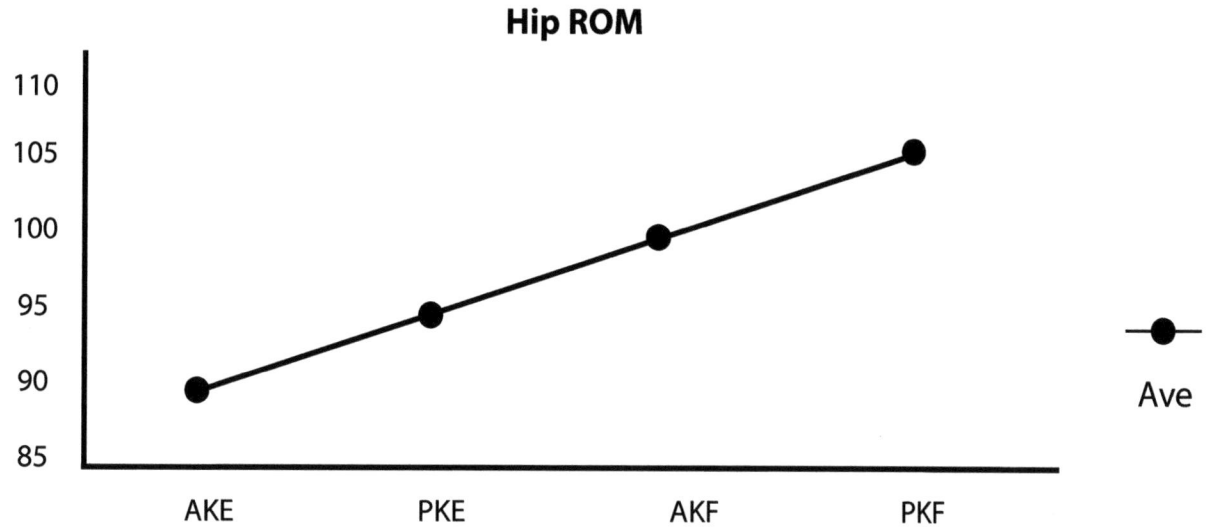

DISCUSSION QUESTIONS

1. Did the class averages fall into the normal range of motion averages? Talk about factors that could affect ROM and flexibility? Give specific examples for each factor.

2. Why would active ROM be less than passive ROM? Why would there be more ROM when the knee is flexed than when the knee is extended regardless of active and passive?

3. What are the pros and cons of active and passive stretching? Is there one that is more recommended? Why?

4. When is stretching recommended, before or after working out? Why?

5. How does lack of hamstring flexibility affect the lumbar spine? People with low back pain are typically recommended to stretch their hamstrings. Why is this usually the case?

Name_____
Section_____ Date_____

6. Among athletes, who is at the most risk for hamstring tears (there may be a few)? Why? What are the risk factors for hamstring injuries?

7. What are the common sites of the hamstrings? Where are tears most common? Why? What are ways that physical trainers rehabilitate athletes with hamstring injuries?

REFERENCES

Anderson, B., and Burke, E. (1991). Scientific, medical, and practical aspects of stretching. *Clinical Journal of Sport Medicine, 10*, 63–86.

Clarkson, H. (2000). Principles and methods. In M. Biblis (Eds.), *Musculoskeletal assessment: Joint range of motion and manual muscle strength* (pp. 4–9). Baltimore, MD: Lippincott Williams and Wilkins.

Floyd, R. T. (2012). *Manual of structural kinesiology* (17th ed.). Dubuque, IA: McGraw-Hill.

Hamilton, N., Weimar, W., and Luttgens, K. (2012). *Kinesiology: Scientific basis of human movement* (12th ed.). New York: McGraw-Hill.

Mitchell, M. (2013). *Introduction to kinesiology: The science of human physical activity.* San Diego, CA: Cognella Academic Publishing.

Roach, K. (1991). Normal hip and knee active range of motion: The relationship to age. *Physical Therapy, 71*, 656–665.

Chapter Four
Exercise Physiology Foundations

CHAPTER ACTIVITY #4

CALORIC INTAKE AND EXPENDITURE

Source: http://web.archive.org/web/20080202213511/http://www.usmc.mil/marinelink/image1.nsf/Lookup/200572675630. Copyright in the Public Domain.

Name_____
Section_____ Date_____

PURPOSE
The purpose of this activity is to review material presented in Chapter 4 of the textbook (Mitchell, 2013) and to better understand the energy balance between caloric intake and expenditure.

INTRODUCTION
As we have learned in Chapter 4 in the text (Mitchell, 2013), all physical activities and physiological functions require energy. Providing this energy are the three nutrients of carbohydrates, fats, and proteins that we consume

Name_____
Section_____ Date_____

from the various foods we eat. In addition, all physical activities and physiology functions expend energy. The amount of caloric intake and expenditure helps determine whether we either gain or lose weight. In this activity, you will estimate your caloric intake and expenditure over a three-day period (preferably three consecutive days) using tables or websites.

CHAPTER SUMMARY QUESTIONS

1. Which of the three nutrients provides the major source of energy for the following activities:

- Running a marathon

  _____

- Lifting a heavy box in one attempt

  _____

- A 50-meter sprint in swimming

  _____

2. Name and describe three physiological functions that change during **acute** exercise:

   - _____

   - _____

   - _____

3. Name and describe three physiological functions that change during **chronic** exercise:

   - _____

   - _____

   - _____

4. Name three negative effects of physical inactivity on the cardiovascular, muscular, or skeletal system. For each, briefly describe how each could hinder performance in some type of physical activity setting (e.g., work, sports, activities of daily living):

   - _____

Name_____

Section_____ Date_____

Name_____
Section_____ Date_____

- _____

- _____

CHAPTER ACTIVITY
Equipment and Materials Needed: Access to the Internet

The purpose of this activity is to understand the energy balance between caloric intake and output for a 72-hour period and see an individual's caloric loss or gain. To maintain an ideal body weight, energy intake (food consumed daily) must equal energy output (daily physical activity). When these two factors are equal, a person is said to be in energy balance. To gain or lose weight, either or both of these factors must be adjusted.

PROCEDURE
1. Select a 72-hour period (normal routine) and measure your nude body weight at the beginning and end of the period.
2. During a 72-hour period, record all food intake. Be specific! Look at the supplement facts. Include grams and calories. Record any types of activity and exercise with time in minutes. Include sleeping time, sitting, standing, walking, everything. When you add the activity time, you should make sure the total minutes in the three 24-hour periods total 4,320 minutes.
3. **Calculate Resting Metabolic Rate (RMR):** The minimum energy expenditure required to carry on normal biologic reactions such as circulation, respiration, and creation and conduction of nerve signals. RMR is approximately 20 to 25 kcal/kg body weight (1 to 1.2 kcal/min) and requires an oxygen consumption of approximately 200 to 250 ml/min. The RMR will be included in the physical activity for 72 hours.

1. **What is energy balance?**
    Caloric balance equation:
    caloric balance (loss or gain) = calories ingested − calories expended

2. **Energy In:** calories from food
    Energy Out: basal metabolism (resting metabolism) + thermal effect of digestion + physical activity

3. **Basal Metabolic Rate (BMR)**
    The minimum energy expenditure required to carry on normal biologic reactions such as circulation, respiration, and creation and conduction of nerve signals (to sustain vital functions in the waking state). BMR is different from resting metabolic rate (RMR) because true BMR is measured in a laboratory setting after complete bed rest, whereas RMR can be measured during any time of day at rest. In adults, BMR is approximately 20–25 kcal/kg body weight and it requires oxygen consumption (VO_2) of about 160–250 ml/kg/min depending on gender, age, body size, or fat free body mass, etc. We will use RMR to estimate caloric expenditure as measuring BMR is not within the scope of this lab.

General formula to calculate RMR:
The **Mifflin-St Jeor equations** are

Name_____
Section_____ Date_____

   Male: BMR = 10 × weight + 6.25 × height − 5 × age + 5
   Female: BMR = 10 × weight + 6.25 × height − 5 × age − 161

These equations require the weight in kilograms, the height in centimeters, and the age in years. To determine your total daily calorie needs, the BMR has to be multiplied by the appropriate activity factor, as follows:

* 1.200 = sedentary (little or no exercise)

* 1.375 = lightly active (light exercise/sports 1–3 days/week)

* 1.550 = moderately active (moderate exercise/sports 3–5 days/week)

* 1.725 = very active (hard exercise/sports 6–7 days a week)

* 1.900 = extra active (very hard exercise/sports and physical job)

NOTE: A 150-pound (68 kg) person walking at 4 miles per hour uses about 300 cal per hour (5 kcal/min). The activity factor **lightly active** corresponds to walking 2 hours per day; **moderately active** corresponds to walking 3 hours per day; **very active** corresponds to walking 4 hours per day; and **extra active** corresponds to walking 5 hours per day (20 miles).

Example calculation of BMR (RMR) and then caloric needs:

25-yo, 6'0" tall, 91-kg male who is moderately active
   RMR: (10*91 kg) + (6.25*182.88 cm) − (5*25 yo) + 5 = 1,933
   Caloric: 1933 * 1.550 = 2,996.15 calories/day

20-yo, 5'5" tall, 45-kg female who is very active
   RMR: (10*45kg) + (6.25*165.10 cm) − 5(20 yo) − 161 = 1220.88
   Caloric: 1220.88 * 1.725 = 2,106.01 calories/day

Another, more generic method:

1) Women: 0.9 kcal/kg body weight/hr
Men: 1.0 kcal/kg body weight/hr
(weight conversion: 1kg = 2.2 lbs)
Ex) 91-kg male
   1.0 kcal/kg/hr* 91kg * 24hr = 2,181 kcal

2) Multiply by 24 hours
Ex) 45-kg female
   0.9 kcal/kg/hr* 45kg * 24hr = 972 kcal

Visit these websites for calories and unit conversion (if you know better ones, use them and cite those instead).
http://caloriecount.about.com/
http://calorieking.com

Name_____
Section_____ Date_____

    http://www.caloriescount.org/calculator.html
    http://www.healthstatus.com/calculate/cbc

If you have a smart phone, a good application to consider using for counting calorie intake and expenditure is *Lose It*.

Name_____

Section_____ Date_____

Data Table 1. Calculations of Caloric Intake

Food	Amount (gram)	Calories (kcal)	Protein (g)	CHO (g)	Fat (g)
Totals					
Convert grams to calories	/////////////	/////////////			
Calculate % of diet	/////////////	/////////////			

Name_____

Section_____ Date_____

Food	Amount (gram)	Calories (kcal)	Protein (g)	CHO (g)	Fat (g)
Totals					
Convert grams to calories	/////////////	/////////////			
Calculate % of diet	/////////////	/////////////			

Name_____
Section_____ Date_____

Food	Amount (gram)	Calories (kcal)	Protein (g)	CHO (g)	Fat (g)
Totals					
Convert grams to calories	/////////////////	/////////////////			
Calculate % of diet	/////////////////	/////////////////			

Name_____
Section_____ Date_____

Data Table 2. Calculation of Caloric Expenditure

Activity (include any kinds)	Performance time (min)	Kcal expended	KJ (if needed)
Total kcal expenditure from **activity**	Total minute should be 24h*60min=**1,440min**		
Kcal expenditure from **BMR**	//////////////////////////////////// ////////////////////////////////////		
Total kcal expenditure	//////////////////////////////////// ////////////////////////////////////		

Name_____
Section_____ Date_____

Activity (include any kinds)	Performance time (min)	Kcal expended	KJ (if needed)
Total kcal expenditure from **activity**	Total minute should be 24h*60min=**1,440min**		
Kcal expenditure from **BMR**	///////////////////////////////// /////////////////////////////////		
Total kcal expenditure	///////////////////////////////// /////////////////////////////////		

Name_____

Name_____
Section_____ Date_____

Activity (include any kinds)	Performance time (min)	Kcal expended	KJ (if needed)
Total kcal expenditure from **activity**	Total minute should be 24h*60min=**1,440min**		
Kcal expenditure from **BMR**	//////////////////////////////// ////////////////////////////////		
Total kcal expenditure	//////////////////////////////// ////////////////////////////////		

REFERENCE

Mitchell, M. (2013). *Introduction to kinesiology: The science of human physical activity*. San Diego, CA: Cognella Academic Publishing.

Name_____
Section_____ Date_____

CHAPTER LAB #4

THE COOPER TEST

Copyright © Chris Brown (CC BY 2.0) at http://commons.wikimedia.org/wiki/File:Marathon_Runners.jpg.

This lab can be modified based on available equipment. The Materials/Methods section has two parts based on equipment used. If blood pressure cuffs are not available, sections within the Introduction, Results, and Discussion will not be relevant.

PURPOSE

To predict and estimate the maximum oxygen consumption (VO_2Max) through a field test, and to investigate the changes in heart rate and blood pressure with exercise.

INTRODUCTION

In general, the utilization of oxygen for demanding activities is directly proportional to the intensity of the exercise (Skinner and McLellan, 1980). As the intensity of the exercise increases, so does the demand for oxygen (Mitchell, 2013, p. 121). Your ability to exercise at moderate to heavy intensities for a prolonged period is referred to as your cardiovascular endurance and the best indicator of this is your VO_2Max (Hoffman, 2009, p. 308).

Skeletal muscles require energy to perform work. The cardiovascular system (CV), which we use for sustained physical activity, is limited by the availability of oxygen. This system is composed of the heart, blood vessels, and blood and is responsible for transporting oxygen to the muscles. The need for oxygen transport to the skeletal muscles increases when we engage in physical activity. The cardiovascular system responds to this demand by increasing blood volume and the distribution of the blood flow during physical activity (Hoffman, 2009, p. 308).

When engaging in physical activity, the amount of blood distributed to the skeletal system increases. At rest, approximately 20% of blood distribution goes to the muscles. With moderate exercise, this increases to approximately 50%. With maximal activity, approximately 70% of our blood is distributed to the muscles (Mitchell, 2013, p. 125).

Name_____
Section_____ Date_____

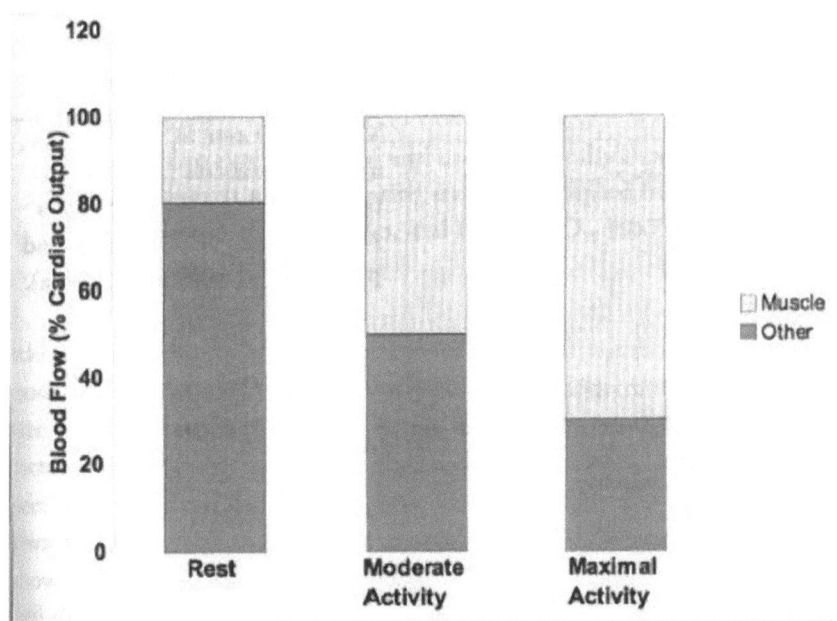

The amount of oxygen that is being delivered to the muscles is also dependent on your cardiac output (CO). Cardiac output is heart rate (HR) x stroke volume (SV). Stroke volume is the amount of blood pumped from the heart by the left ventricle in each heart beat, and heart rate is the number of heart beats in one minute. This means that cardiac output is the amount of blood pumped by the heart in one minute (Mitchell, 2013, p. 123). The amount of oxygen delivered to the tissues depends on how much oxygen is in the blood and how much blood the heart is pumping. During physical activity both heart rate and stroke volume increase. Therefore, cardiac output increases as muscles use more oxygen (Mitchell, 2013, pp. 121–3). Oxygen uptake (VO_2), or the amount of oxygen used by the tissues, increases in direct proportion to the intensity of exercise until maximal oxygen (VO_2Max) is reached (Hoffman, 2009, p. 308). At this point, oxygen consumption plateaus regardless of continued exercise.

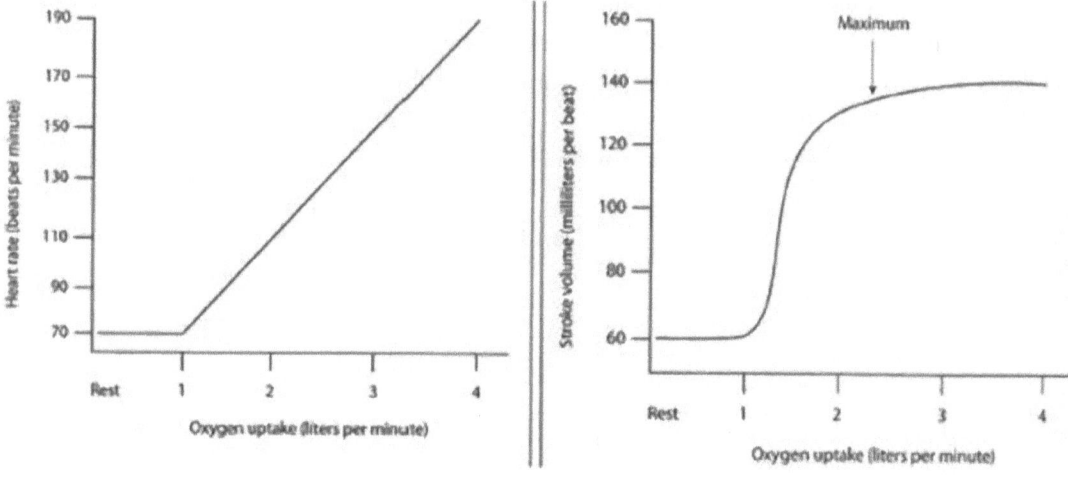

In both stroke volume and heart rate, a maximum will be reached. The heart can only pump so much blood out in one beat at any given time. Stroke volume plateaus early on in physical activity but oxygen consumption continues to increase. This is because your heart rate is still increasing (Mitchell, 2013, pp. 123–4). Your heart rate, however, will only be able to increase until it reaches it maximum rate. At this point, you have reached your maximum cardiac output as well as your VO_2Max (Hoffman, 2009, p. 308).

Name_____
Section_____ Date_____

In addition to changes in HR, CO, and VO_2 during physical activity, there also are changes in blood pressure (BP). Blood pressure is related to CO and to total peripheral resistance (TPR) in the blood vessels. Total peripheral resistance is the sum total of vascular resistance to the flow of blood in the systemic circulation.

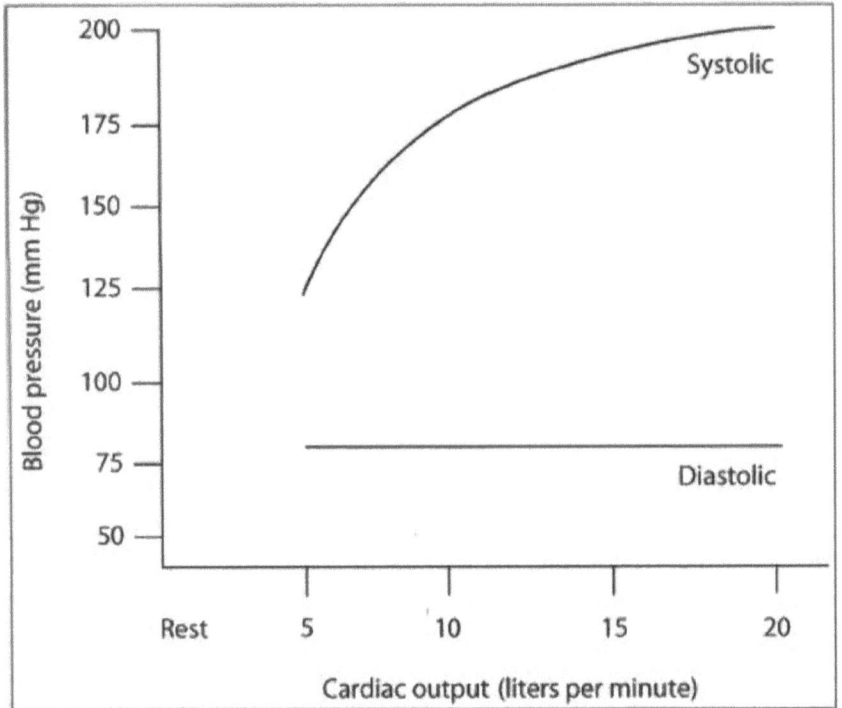

With exercise, CO increases and TPR generally decreases, resulting in a large increase in blood pressure. The increase in cardiac output causes a direct increase in systolic blood pressure but very little change in diastolic (Mitchell, 2013, p. 124).

This lab experiment will use an indirect method for estimating VO_2Max. The Cooper Test in a field test that estimates VO_2Max with reasonable accuracy and was developed by Dr. Kenneth Cooper. Dr. Cooper generated the following chart as a means of estimating VO_2Max without the need for equipment.

Twelve-Minute Run/Walk and Oxygen Consumption

Distance (miles)	1/4 Mile Laps	VO_2 MAX	Classification
less than 1	less than 4	25	
1	4	25	Low (VO_2 = 30–35)
1.065	4.25	27	
1.125	4.5	29	
1.187	4.75	31.6	
1.25	5	33.8	
1.317	5.25	36.2	Fair (VO_2 = 36–40)
1.375	5.5	38.2	
1.437	5.75	40.4	Good (VO_2 = 41–45)

Name_____
Section_____ Date_____

1.5	6	42.6	
Distance (miles)	**1/4 Mile Laps**	**VO$_2$ MAX**	**Classification**
1.565	6.25	45	**Very Good**
1.625	6.5	47.2	**(VO$_2$ = 46–50)**
1.687	6.75	49.2	**Excellent**
1.75	7	51.6	**(VO$_2$ = 51)**
1.817	7.25	53.8	**Superior**
1.875	7.5	56	**(VO$_2$ = 55)**
1.937	7.75	58.2	**Competitive**
2	8	60.2	**(VO$_2$ = 60)**

Note: If you do not fall under the given categories, you will use this equation to approximate your VO$_2$Max:

$$VO_2 max\ (ml/kg/min) = (D\ in\ miles - 0.3138)/0.0278$$

MATERIALS/METHODS

Participants

Your fellow classmates are your participants.

Task and Apparatus Part 1

- Blood pressure cuff (digital automatic on the wrist)
- Stopwatch
- 400-meter track
- Athletic attire

Task and Apparatus Part 2

- Stopwatch
- 400-meter track
- Athletic attire

Procedure Part 1

The researchers will be running the experiment on the rest of the participants. Have the students put on the blood pressure cuffs and keep them on for the entire procedure. Prior to the start of the run, the participants will take and record their heart rate and blood pressure. Both of these measures can be collected with the blood pressure cuff. If BP cuff has a memory device, have the students store the data after each recording. They will then warm up for two minutes by walking or jogging around the track. Heart rate and blood pressure will then be taken again. The participants will then jog or run around the track for 12 minutes. After such time, heart rate and blood pressure will be taken again. The participants will also record how many laps they completed in the 12 minutes. The researchers will collect all the data and average the heart rates and blood pressures at the three different times.

Name_____
Section_____ Date_____

Procedure Part 2

The researchers will be running the experiment on the rest of the participants. Prior to the start of the run, the participants will take and record their heart rates. Heart rate will be taken by counting the number of beats felt in 15 seconds then multiplying by four. They will then warm up for two minutes by walking or jogging around the track. Heart rate will then be taken again. The participants will then jog or run around the track for 12 minutes. After such time, heart rate will be taken again. The participants will also record how many laps they completed in the 12 minutes. The researchers will collect all the data and average the heart rates at the three different times.

RESULTS

Subject Info

	Heart Rate	Blood Pressure	Laps Completed	Predicted VO$_2$Max
Pre-Warm-Up (Time 1)				
Post-Warm-Up (Time 2)				
Post-Run (Time 3)				

Averages for HR overtime

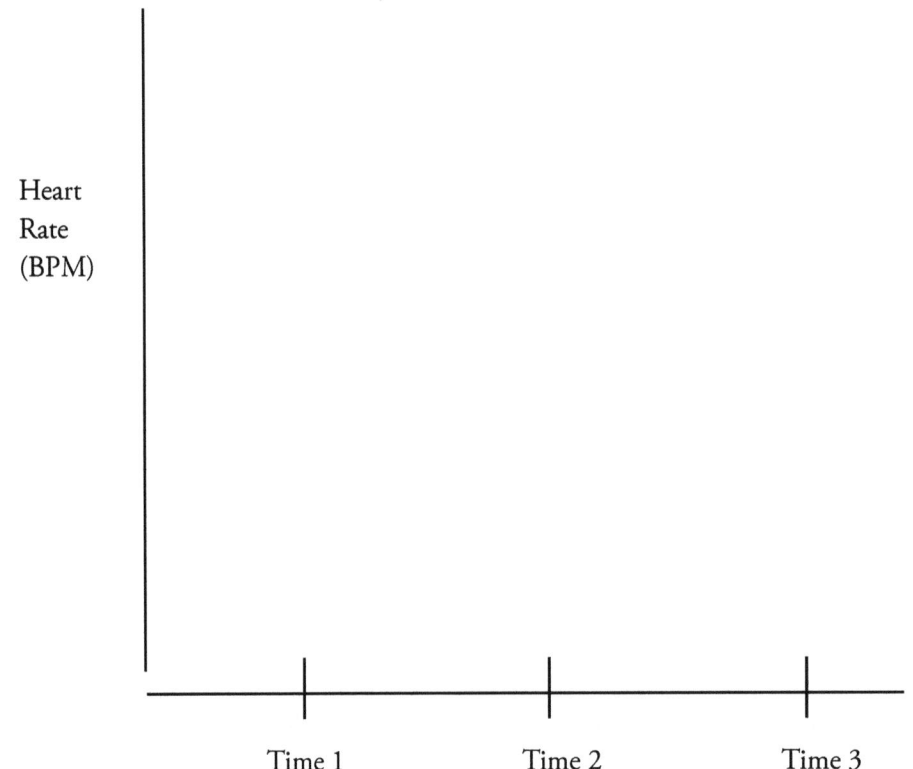

58 Introduction to Kinesiology: The Science of Human Activity - Lab Manual

Name_____
Section_____ Date_____

Averages for BP overtime

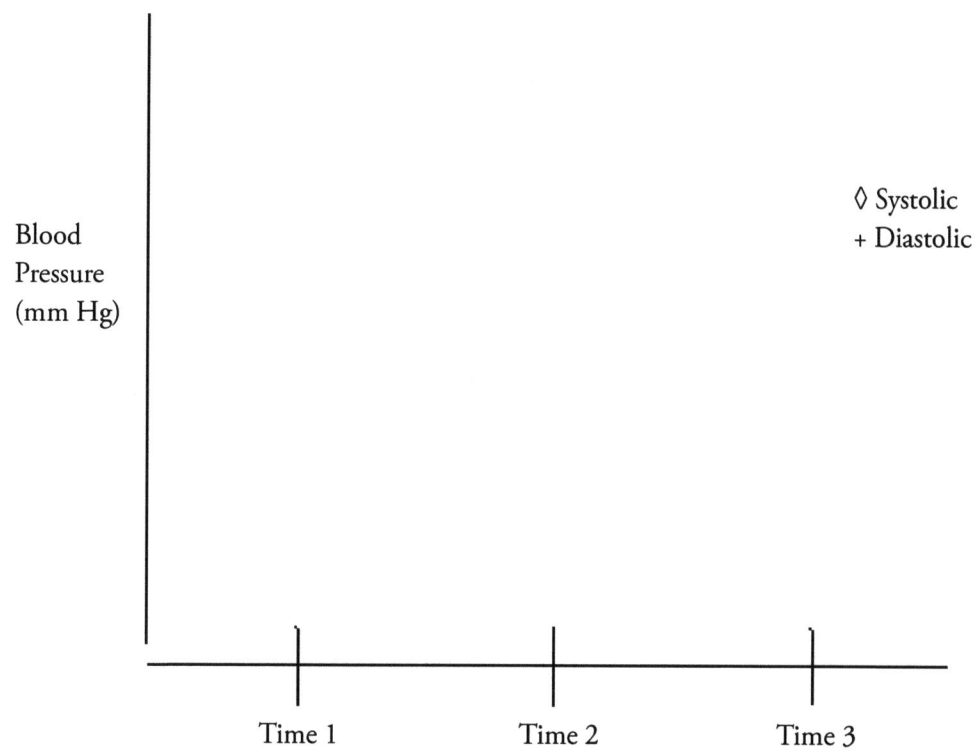

Blood Pressure (mm Hg)

◊ Systolic
+ Diastolic

Time 1 Time 2 Time 3

Heart rate responses of fit and unfit subjects

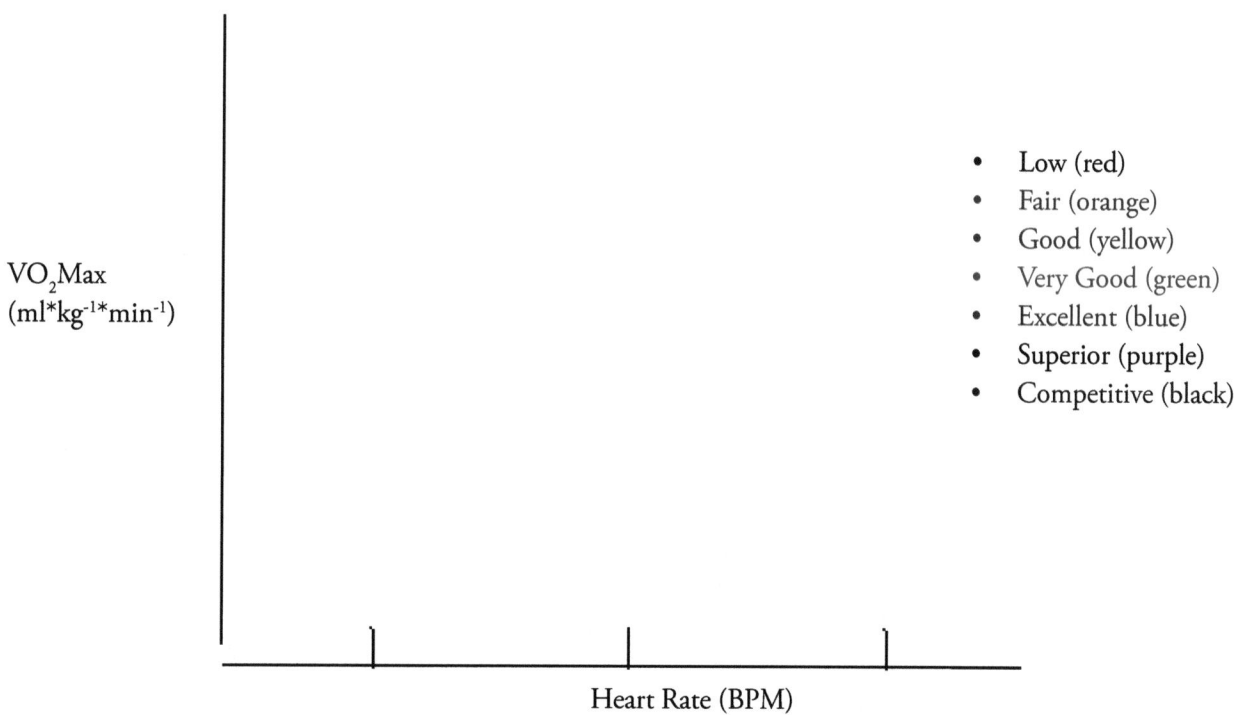

VO_2Max ($ml*kg^{-1}*min^{-1}$)

- Low (red)
- Fair (orange)
- Good (yellow)
- Very Good (green)
- Excellent (blue)
- Superior (purple)
- Competitive (black)

Heart Rate (BPM)
BPM, BP, VO_2 graphs from Mitchell (2013) Cognella Academic Publishing.

Name_____

Section_____ Date_____

Chapter Five
Biomechanical Foundations

DISCUSSION QUESTIONS

1. What are the effects of endurance training on CO? Explain this in terms of changes in HR and SV. Why do these changes occur?

2. Why does only systolic BP change and not diastolic? How does endurance training affect overall blood pressure at rest and with exercise? What might be the reason for this?

3. How does the body change the distribution of blood flow to the muscles? Explain why there is a need for increased blood flow for the muscles to work during physical activity.

4. Explain the difference in CO, HR, and VO_2Max between fit and unfit individuals while engaging in physical activity.

5. When one starts exercising, there is an immediate increase demand for O_2. Can the CV system meet this need immediately? Explain the phases in O_2 consumption versus O_2 requirement when exercising.

6. How does VO_2Max change as we age? What are some of the reasons for this? What are some ways to slow this process?

7. What is the gold standard for VO_2Max testing? Explain how this test works.

Name_____

Section_____ Date_____

REFERENCES

Hoffman, S. (2009). *Introduction to kinesiology: Studying physical activity*. Champaign, IL: Human Kinetics.

Mitchell, M. (2013). *Introduction to kinesiology: The science of human physical activity.* San Diego, CA: Cognella Academic Publishing.

Skinner, J., and McLellan, T. (1980). The transition from aerobic to anaerobic metabolism. *Research Quarterly for Exercise and Sport, 51,* 234–48.

Name_____
Section_____ Date_____

CHAPTER ACTIVITY #5

BIOMECHANICAL ANALYSIS OF TOOLS

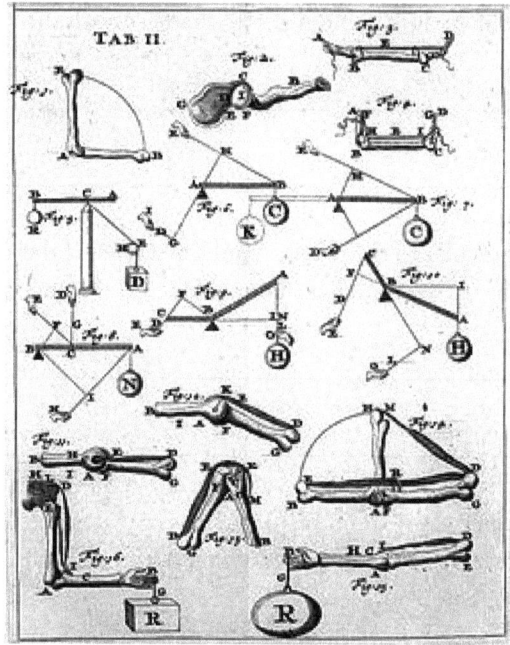

Source: Giovanni Alfonso Borelli, De Motu Animalium. Copyright in the Public Domain.

PURPOSE

The purpose of this activity is to review material presented in Chapter 5 of the textbook (Mitchell, 2013) and to better understand the concepts of torque and mechanical advantage by analyzing how different tools produce angular force.

INTRODUCTION

In Chapter 5 of the text (Mitchell, 2013), we examined biomechanical principles of human movement. One of the important concepts was the property called torque, the force that an object can produce in angular motion. Because angular motion is a very common type of movement produced by the human body, torque is a very important element to understand. In various skills and activity, we often use tools to help tasks. In this activity we review some of the concepts in Chapter 5, and also explore certain biomechanical principles involved in different tools.

CHAPTER SUMMARY QUESTIONS

1. A group of children decide to play a game on a teeter-totter that is 3 m in length. One of the children, weighing 30 kg, sits on one end of the teeter-totter. The game is to require each of the other children to take turns sitting somewhere along the other end of the teeter-totter to obtain perfect balance. If one of the other children weighs 45 kg, where would that child have to sit on the teeter-totter to obtain perfect balance with the first child? (Hint: Use the principle of levers.)

Name_____
Section_____ Date_____

_____

_____

2. Could any child weighing less than 30 kg obtain perfect balance with the first child? Explain your answer:

_____

_____

3. A swimmer is oriented perpendicular to the parallel banks of a river. If the swimmer's velocity is 2 m/s and the current is 0.5 m/s, what will be the swimmer's resultant velocity? (Hint: Use Pythagorus's theorem.)

_____

_____

4. The relative angle at the knee changes from 180 degrees to 95 degrees during the knee flexion phase of a squat exercise. If 10 complete squats are performed, what is the total angular distance and the total angular displacement undergone at the knee?

_____

5. How much force must be applied by a kicker to give a stationary 2.5 kg ball an acceleration of 40 m/s^2? (Hint: Use one of Newton's laws.)

_____

CHAPTER ACTIVITY
Equipment and Materials Needed: Access to the Internet

Go to the following websites for the three tools shown below. For each tool, briefly describe its history (who developed it, and when). Then describe one way that you could increase the torque of the device (i.e., its ability to produce angular force). In your answer refer to the formulas for torque and/or mechanical advantage.

Source: http://upload.wikimedia.org/wikipedia/commons/archive/8/84/20100909225652!Claw-hammer.jpg. Copyright in the Public Domain.

Name_____
Section_____ Date_____

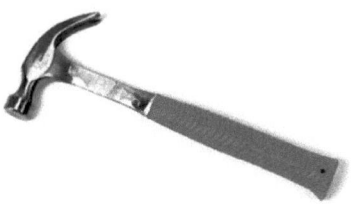

Brief history:

How can torque be increased by modifying this tool?

Source: http://en.wikipedia.org/wiki/File:Tool-pliers.jpg. Copyright in the Public Domain.

Brief history:

How can torque be increased by modifying this tool?

Copyright © Fcb981 (CC BY-SA 3.0) at http://en.wikipedia.org/wiki/File:Screw_Driver_display.jpg.

Biomechanical Foundations 65

Name_____
Section_____ Date_____

Brief history:

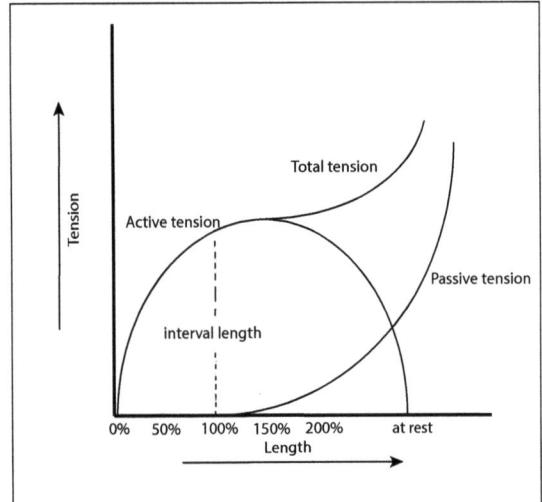

Passive elastic tension in a muscle passively stretched to increasing lengths. (Line originating at resting length below the dotted line.)

Total tension exerted by the muscle (the top line).

Active tension is total tension minus passive tension. It is the muscle actively contracting. (Hamilton, Weimar, and Luttgens, 2012, pp. 52–53) (The line resembling an inverted "U.")

Adapted from: R.T. Floyd, *Manual of Structural Kinesiology*. Copyright © 2012 by McGraw Hill.

How can torque be increased by modifying this tool?

REFERENCE

Mitchell, M. (2013). *Introduction to kinesiology: The science of human physical activity.* San Diego, CA: Cognella Academic Publishing.

Name_____
Section_____ Date_____

CHAPTER LAB #5.1

LENGTH–TENSION RELATIONSHIP IN MUSCLES

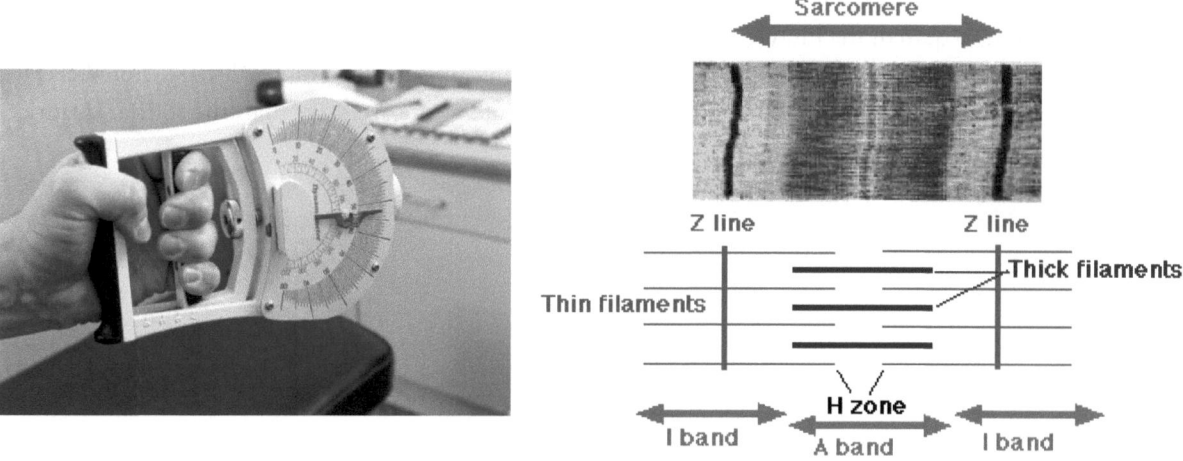

Copyright © 2010 by iStockphoto.com/BanksPhotos.

Copyright © Sameerb at http://commons.wikimedia.org/wiki/File:Sarcomere.gif.

PURPOSE
The purpose of this experiment to investigate the length–tension relationship in muscles. (This lab could also be done in Chapters 2 or 6.)

Length-tension curves for isolated muscles

INTRODUCTION
During this lab session, we will examine the effect of joint angle on a muscle's force production. When a muscle is maximally contracted, it is at its shortest length. When a muscle is fully stretched, the muscle is at its longest length. When a muscle is neither short nor long, it is at what is called its resting length (Mitchell, 2013, p. 82). But you can also change the muscle length by simply changing the joint angle. By changing the muscle length, the amount of force the muscle can generate changes as well. How short or long the muscle is when asked to contract (isometrically, concentrically, or eccentrically) affects the amount of force it will produce. This is called a length–tension relationship (Floyd, 2012, p. 58).

EXPERIMENT 1: THE EFFECT OF JOINT ANGLE ON GRIP FORCE

Tension in a muscle can be thought of as a pulling force. Tension may be either active or passive. Passive tension is developed as a muscle is stretched past its resting length. Passive tension is like a rubber band; as the length increases, so does the passive tension. Active tension in the muscle is dependent on the amount of motor units and the respective muscle fibers recruited in the contraction (Floyd, 2012, p. 58).

There is an optimum length at which a muscle, when stimulated, can exert maximum tension. This length varies somewhat according to both the muscle's structure and its function (Hamilton et al., 2012, pp. 52–53). Typically, a

Name_____

Section_____ Date_____

muscle can generate the most active tension around 130% of resting length where resting length is 100% of muscle length. As the muscle is stretched beyond this length, the amount of active tension the muscle can generate decreases significantly. Likewise, when the muscle is shortened, its ability to generate active tension decreases as well. When a muscle is shortened to around 50% to 60% of its resting length, its ability to develop contractile tension is essentially reduced to zero. Passive tension begins to increase when the muscle is lengthened past its resting length (100%) and continues to increase as the muscle continues to be stretched (Floyd, 2012, p. 58). This relationship suggests that when maximum force is required, the muscle should be longer than resting length. This relationship applies for all three contraction types. It should also be noted that a longer tendon can generate a higher level of stored elasticity than a shorter tendon. For example, the Achilles tendon can generate greater velocity than the shorter quadriceps tendon (Hamilton et al., 2012, pp. 52–53).

Name_____
Section_____ Date_____

MATERIALS/METHODS

Participants

The participants will be your fellow class members.

Task and Apparatus

- Hand-held dynamometer
- Goniometer

Procedure

Divide the class into groups, with one researcher designated for each group. This person will ensure that the protocol is followed:

Using the hand-held dynamometer and the goniometer, complete the following chart and graph your results. Place your nondominant arm by your side with your elbow straight. While holding the dynamometer, start with the wrist at 60 degrees of flexion (measure using the goniometer) and perform a maximal isometric contraction of the wrist flexors. An isometric contraction is where the muscle fibers contract but the limb does not move (Mitchell, 2013, p. 85). Record the grip force in the table for the rest of the wrist angles. In all the cases, use the goniometer to measure the wrist angle and perform a maximal isometric contraction of the wrist flexors.

DATA ANALYSIS

Wrist Angle	Grip Force
60 degrees of wrist flexion	
45 degrees of wrist flexion	
30 degrees of wrist flexion	
15 degrees of wrist flexion	
Wrist neutral	
15 degrees of wrist extension	
30 degrees of wrist extension	
45 degrees of wrist extension	
60 degrees of wrist extension	

Name_____

Section_____ Date_____

Name_____
Section_____ Date_____

Average the entire class's results and graph:

Force

Flexion	0	Extension

Length

Name_____
Section_____ Date_____

DISCUSSION QUESTIONS

1. Explain the structure of muscle. Start with the epimysium and finish with the sarcomere. What is a sarcomere? Explain the structure of the sarcomere and how actin and myosin cause a contraction.

2. How do the cross-bridges affect the force–length curve? Is there an ideal length for these cross-bridges to bind? If so, what is it? Explain why.

3. What are the active forces in muscle and what are the passive forces? How are these passive forces contributing to overall force production?

4. What does resting length mean in terms of a muscle's length? Approximately where is resting length for these muscles (degrees)?

5. At what wrist angle was the most force produced? Did your results (graph) seem similar to the active tension curve? Explain your results by comparing/relating to the curve.

6. Think about how this is relationship could be used in performing a motion for maximal contraction. Give specific examples.

REFERENCES

Hamilton, N., Weimar, W., and Luttgens, K. (2012). *Kinesiology: Scientific basis of human movement* (12th ed.). New York: McGraw-Hill.

Floyd, R. T. (2012). *Manual of structural kinesiology* (17th ed.). New York: McGraw-Hill.

Mitchell, M. (2013). *Introduction to kinesiology: The science of human physical activity.* San Diego, CA: Cognella Academic Publishing.

Name_____
Section_____ Date_____

CHAPTER LAB #5.2

THE PHYSICS OF STABILITY

Copyright © Xeaza (CC BY-SA 3.0) at http://commons.wikimedia.org/wiki/File:SmirkusRolaBolaHandBalance.jpg.

PURPOSE
To examine how resistance to angular motion influences motor performance.

INTRODUCTION
An object's inertia is a measure of it resistance to linear motion. As we discussed in Chapter 5 (Mitchell, 2013), Newton's First Law of Motion states that an object will remain at rest or in a state of constant motion unless acted on by an external force. An object's *moment of inertia* is a measure of its resistance to angular motion (rotation). *Moment of inertia* is a function of the object's mass and the way that mass is distributed from the axis about which it rotates. For example, if you held a baseball bat at the handle and *played* the bat around by moving your wrists, the bat's resistance to angular motion would be higher than if you made the same movements but held the bat by the other end. Although the mass of the bat does not change, more mass is distributed farther from the axis of rotation when the bat is held by its handle than when held by its other end. Consequently, its *moment of inertia* is higher when held by the handle.

We often change the whole body's moment of inertia or a body segment's moment of inertia to speed up or slow down a movement or make a movement more or less effortful. For example, an ice skater can spin faster about the longitudinal axis when the arms are tucked close to the body than when held away from the body. Similarly, a gymnast can spin faster about the transverse axis when the legs, arms, head, and neck are tucked inward than when extended. Bending the leg during the swing phase in gait reduces the leg's moment of inertia and permits it to swing faster and with less effort. The same effect occurs when we bend the arm during the recovery portion of the front crawl stroke in swimming.

Whole body moment of inertia also has an influence on postural control. For example, a child will oscillate more quickly when standing upright than an adult because the child has a lower whole body moment of inertia relative to the ankle joint. Consequently, different body shapes and sizes place different demands on postural control mechanisms. The same principle applies when we attempt to balance objects that vary in shape and size.

In this lab, you will analyze how changing the moment of inertia of a stick influences how difficult it is to balance that stick.

Name_____

Section_____ Date_____

Name_____
Section_____ Date_____

MATERIALS/METHOD

Participants

Everyone in the class should participate in this experiment. The class will be paired up for data collection.

Task and Apparatus

The task is to balance a stick on the index finger for as long as possible. Three different sticks will be used (30 cm, 50 cm, and 100 cm). The ideal stick is a TV antennae because it can be lengthened and shortened (i.e., the moment of inertia can be changed), yet the mass of the stick stays the same.

Each trial with each stick will last 5 minutes (though the time can be shortened). The end of the stick should be positioned on the tip of the index finger (either hand) and held vertically initially. When the trial begins, the performer must immediately release the stick and attempt to balance it. If the stick falls from balance or touches any part of the person's body during balancing, the stick must be repositioned immediately and another attempt to balance the stick must be started. If the stick falls to the ground, 5 seconds should be added to the trial. The **dependent variable** is the number of times the stick falls from balance during the 5-minute trial.

Procedure

The class is divided into pairs, with the pairs flipping a coin to determine who will perform the task first and who will record data first. The first performer will randomly choose one stick length to balance, and the recorder will keep time and document the number of times the stick falls from balance. After the first 5-minute trial, the pairs will swap roles. The recorder will randomly choose a stick length to balance (it should be different from the length chosen by the first performer) and complete his or her 5-minute trial. The first performer times the trial and records performance. The pair continues to alternate between performer and recorder until each person has balanced each of the sticks for 5 minutes (15 minutes total).

Name_____
Section_____ Date_____

Data Analysis

Record the scores for each stick in the table below.

	Falls from balance in 5 min
Long stick (100 cm)	
Medium stick (50 cm)	
Short stick (30 cm)	

RESULTS

Use the bar graph below to plot your individual data. Combine all participants' data together and compare how well each of the sticks could be balanced.

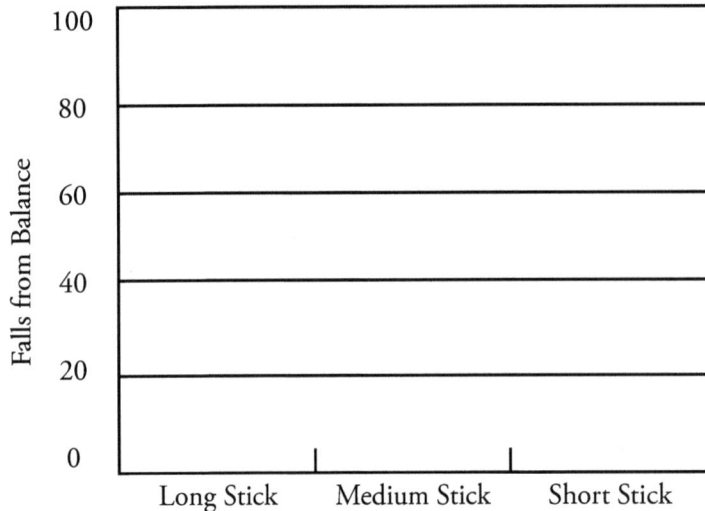

DISCUSSION QUESTIONS

1. Which was the easiest stick to balance and which was the most difficult?

2. What strategies did you use to balance the sticks?

3. Did you use the same strategies for each stick? If not, why?

4. Why do you think the sticks varied in their difficulty to balance?

5. Describe situations where someone might modify the moment of inertia of the whole body or a body segment to increase or decrease stability.

6. How might changes in body mass and body proportions during growth contribute to changes in postural stability?

REFERENCE

Mitchell, M. (2013). *Introduction to kinesiology: The science of human physical activity.* San Diego, CA: Cognella Academic Publishing.

Chapter Six
Motor Control and Motor Learning Foundations

CHAPTER ACTIVITY #6

INSTRUCTIONS AND OBSERVATION IN LEARNING A COMPLEX MOTOR SKILL

Copyright © James Heilman at http://en.wikipedia.org/wiki/File:5_ball_juggling.jpg.

PURPOSE
The purpose of this activity is to review material presented in Chapter 6 of the textbook (Mitchell, 2013) and to determine the effectiveness of observation with and without instruction in learning a complex motor skill.

INTRODUCTION
In Chapter 6, we explored many of the concepts and principles of motor control and motor learning. Some of these concepts are related to the many factors that help you in learning a new motor skill. The information you receive about the skill itself, and the feedback you receive about how you are performing, are important sources of information that can help you learn a new motor skill. One type of information that can be used to help you learn how to do a motor skill is called a *demonstration* or *observational* learning (McCullagh and Weiss, 2002). Typically the learner observes someone, either live or by watching a video, perform the skill to be learned, and then the learner attempts

Name_____
Section_____ Date_____

to perform the skill following the demonstration. This activity will allow you to explore these types of information that might be helpful in learning a complex motor skill.

CHAPTER SUMMARY QUESTIONS

1. A person decides to reach forward and pick up a coffee cup. At each level of the neuromuscular system (higher center level, spinal level, and lower level), describe five neuromuscular structures that would be involved from the planning through the execution of the movement. List these in chronological order:

    - _____

    - _____

    - _____

    - _____

    - _____

2. For this same movement, name four joints that are involved in this movement (Hint: Refer to Chapter 3 for help):

    - _____

    - _____

    - _____

    - _____

3. How many degrees of freedom of joint movement can you count for this movement? Justify your answer:

    Total degrees of freedom: _____

    Justify you answer: _____

    _____

    _____

4. Identify one model of motor control. Using this model, describe how catching a ball is accomplished.

    Model: _____

Name_____

Section_____ Date_____

Description of catching: _____

_____

_____

_____

5. Pick one type of sport skill that you are trying to teach someone. List and briefly describe three types of feedback you could give to the learner:

- _____

- _____

- _____

CHAPTER ACTIVITY

Equipment and Materials Needed: Access to the Internet, three tennis balls or bean bags, stopwatch or timer

In this activity, you will be trying to learn how to juggle three balls or objects. A good place to juggle is next to your bed, and standing so that if you drop a ball (balls) it will fall on the bed and be easily retrievable. In this activity, you will perform in three conditions.

Condition 1—You will read instructions on how to juggle one ball, two balls, and finally three balls. After reading the instructions you will attempt to duplicate what you have read.

Condition 2—You will watch video clips of someone juggling one, two, and finally three balls. The videos contain audio instructions but **you will turn the audio off**, and only watch the person demonstrating.

Condition 3—You will watch videos of someone juggling one, two, and finally three balls. **You will turn the audio on** so that you can watch the videos **and** also listen to the verbal instructions.

Before you start, you need to determine the order of the conditions that you will perform under. Below are the possible orders:

Order 1: Condition 1, Condition 2, Condition 3
Order 2: Condition 1, Condition 3, Condition 2
Order 3: Condition 2, Condition 3, Condition 1
Order 4: Condition 2, Condition 1, Condition 3
Order 5: Condition 3, Condition 1, Condition 2
Order 6: Condition 3, Condition 2, Condition 1

Name_____
Section_____ Date_____

    To determine the order, cut up six small but identical-size pieces of paper and write each order on one of the six pieces of paper. Put the six pieces of paper containing all six orders in a bowl, stir up the pieces of paper, close your eyes, and pick out one piece of paper from the bowl. This is the order that you will do the activity.

----A NOTE TO INSTRUCTORS----
Because the above procedure of ordering the conditions is randomized across students, there should be approximately an equal number of students in each of the orders. You may wish to get an average score for all six orders to make comparisons across the various conditions.

In the space below, fill in which order you used in this activity:

Condition 1—Written Instructions Only

One-Ball Juggling

First, start by holding one ball in your hand, palms up, with both hands at about waist level. Throw the ball up, a little higher than your head and in an arch, and catch it with your other hand. Repeat by starting with the other hand. Start slow and speed up a little so that the ball only stays in each hand for a moment. Keep your eye on the ball at all times.

Two-Ball Juggling

Now, hold one ball in each hand. Throw only one of the balls as you did before, back and forth, but hold on to the other ball in one of your hands.

    Once you are comfortable with that, now try throwing one of the balls to the other hand, and when this ball reaches its peak, throw the other ball to the other hand as before. Start slow, making sure you catch each ball before starting over. You might try throwing the balls a little higher to give yourself more time to catch the ball. As you get comfortable, lower the height of the throws. Keep your eyes looking up at the balls, not at your hands. Look at the first ball as it reaches its peak, then switch your vision to look at the second ball at its peak.

    Once you are comfortable throwing the balls in one direction, switch to the other direction. Initially, you may have started throwing with the left hand. Now start by throwing with the right hand.

    Once you are comfortable juggling two balls, now try speeding it up.

Three-Ball Juggling

For three-ball juggling, hold two balls in one hand and one ball in the other. Begin by tossing one of the balls in the hand that is holding two balls, to the other hand. As in two-ball juggling, toss the ball in the other hand back to the hand that was holding two balls. In other words, juggle just two balls, but hold the third ball in one of the hands.

Name_____
Section_____ Date_____

It's a little easier if you toss the balls a little higher and at a moderate speed. As you get more comfortable, you can lower the height of the balls, and increase the speed.

Now it is time to add the third ball. As before, toss the first ball, and when it reaches its peak, toss the second ball. When the second ball reaches its peak, toss the third ball. If successful, you should have tossed all three balls without dropping any of them. So, try juggling all three balls, but stop after catching the third ball. If you can do this three times in a row, it is time to start your activity.

You will need a stopwatch or some type of timing device—perhaps an application on your smart phone—that allows you to time a one-minute trial. A good place to juggle is next to your bed, and standing so that if you drop a ball (balls) it will fall on the bed and be easily retrievable. You will try to juggle the three balls after you have viewed the video clips above. You will perform 10 trials and record the number of balls successfully juggled during a one-minute trial. To measure the number of balls successfully juggled, you will simply count how many times you toss a ball up with either hand during the trial.

To begin a trial, first start the timer. Then, begin juggling.

Use the table below to record your results.

Trial	# of balls
1	
2	
3	
4	
5	
6	
7	
8	
9	
10	

Now, obtain the mean or average number of balls you successfully juggled by adding the total number of balls tossed and dividing the total by 10.

Put the mean in the box below:

Your mean of 10 trials is:

Condition 2—Observation Without Audio Instructions

Go to the following website and listen to the introduction:

Name_____
Section_____ Date_____

http://www.thejimshow.com/juggle/index.html

Scroll down to "Juggling Introduction" and click on the large screen version for either Quicktime or Windows.

After watching and listening to the introduction, go to the following video clips and watch them but make you sure **you turn off the volume**. Try to get as much information about juggling one, two, and then three balls by just watching the video clip, but not listening to them:

Watch these video clips in the following order:

Juggling Step #1
Juggling Step #2
Juggling Step #3

After watching the above video clips you are now ready to perform 10 trials performing three-ball juggling.

To begin a trial, start the timer. One trial is equal to one minute. Begin juggling.

Use the table below to record your results.

Trial	# of balls
1	
2	
3	
4	
5	
6	
7	
8	
9	
10	

Now, obtain the mean or average number of balls you successfully juggled by adding the total number of balls tossed and dividing the total by 10.

Put the mean in the box below:

Your mean of 10 trials is:

Condition 3—Observation with Audio Instruction

Go to the following website:

Motor Control and Motor Learning Foundations   83

Name_____

Section_____ Date_____

http://www.thejimshow.com/juggle/index.html

Scroll down to "Juggling Introduction" and click on the large screen version for either Quicktime or Windows.

After watching and listening to the introduction, go to the following video clips and watch and listen to them. Try to get as much information about juggling one, two, and then three balls.

Watch these video clips in the following order:

Juggling Step #1

Juggling Step #2

Juggling Step #3

After watching the above video clips you are now ready to perform 10 trials performing three-ball juggling. To begin a trial, start the timer. One trial is equal to one minute. Begin juggling.

Name_____
Section_____ Date_____

Use the table below to record your results.

Trial	# of balls
1	
2	
3	
4	
5	
6	
7	
8	
9	
10	

As before, obtain the mean or average number of balls you successfully tossed by adding the total number of balls tossed and dividing the total by 10.

Put the mean in the box below:

Your mean of 10 trials is:

DISCUSSION QUESTIONS

Compare the mean of your performance under the three conditions: 1) after only verbal instructions, 2) after only watching the video, and 3) after watching the video that also included verbal instructions.

1. Which condition resulted in the best mean performance?

2. What types of information were the most important in the videos that you watched?

3. What types of information were not particularly helpful in the videos?

4. Did you think the order of performing the three conditions affected your learning? If so, in what way?

Name_____

Section_____ Date_____

REFERENCES

McCullagh, P., and Weiss, M. R. (2002). Observational learning. The forgotten psychological method in sport psychology. In Van Raalte, J. L. and Brewer, B. W. (Eds.). *Exploring sport and exercise psychology*. (2nd ed. 131–150). Washington, DC: American Psychologist Association.

Mitchell, M. (2013). *Introduction to kinesiology: The science of human physical activity*. San Diego, CA: Cognella Academic Publishing.

www.thejimshow.com/juggling

Name_____
Section_____ Date_____

CHAPTER LAB #6.1

HICK-HYMAN LAW: SIMPLE VS. CHOICE REACTION TIME

Source: http://commons.wikimedia.org/wiki/File:200mW_at_Josef_Odlozil_Memorial_in_Prague_14June2010_065.jpg. Copyright in the Public Domain.

PURPOSE

To investigate the principles of Hick-Hyman Law (simple vs. choice reaction time) by estimating reaction time in the following experiment.

INTRODUCTION

The principle that the motor control system requires preparation before initiation of an action is known as *action preparation* (Magill, 2010). This principle is derived from the various factors in observed differences in the amount of time between the onset of the stimulus and the onset of the movement (Mitchell, 2013). This time interval is known as reaction time (RT). RT is able to tell us that preparation to produce voluntary movement takes time. Certain actions and circumstances require more time than others for preparation (Magill, 2010).

The number of response alternatives the performer has to choose from in both task and performance situations is an important factor that influences preparation time. As the number of choices increase, the amount of preparation time increases as well. RT is the fastest when there is only one choice response to one stimulus (simple reaction time). RT slows down when more than one stimulus and more than one response are possible (choice reaction time). This relationship between the number of choice responses and the amount of reaction time increase is so stable that it is known as Hick-Hyman Law (Magill, 2010; Mitchell, 2013).

Hick (1952) and Hyman (1953) investigated the relationship between the number of stimulus-response alternatives and choice reaction time using various numbers of lights with an equal number of keys that were to be pressed when the appropriate light lit. They discovered that the choice reaction time appeared to increase by a nearly constant amount (about 150 ms) every time the number of stimulus-response alternatives doubled. They were able to conclude that there is a linear relationship between the choice RT and the logarithm of the number of choice-response alternatives (Schmidt and Lee, 2011; Mitchell, 2013). This means that if we calculate the logarithm of the

Name_____
Section_____ Date_____

number of choices in a choice RT situation and plot the resulting graph, RT should increase linearly as the number of choice-response alternatives increase (Magill, 2010). The equation that describes this law is:

RT = K $\log_2$ (N + 1), where K in a constant (simple RT) and N equals the number of choice responses. This relationship when graphed looks like the following:

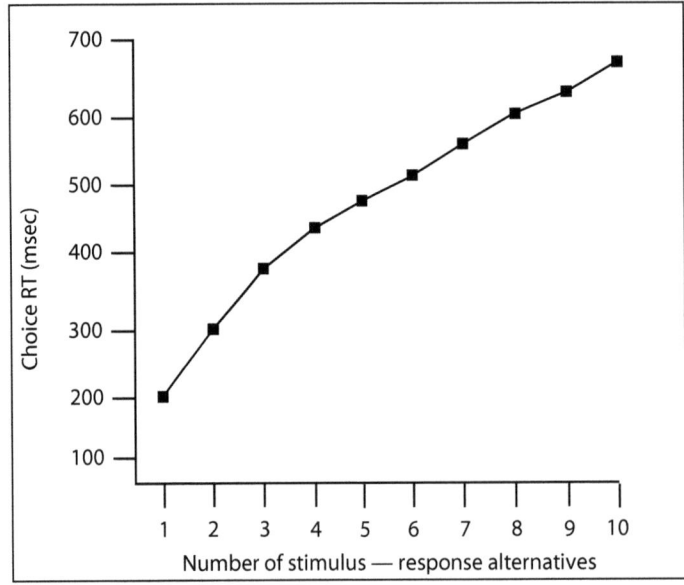

Predicted reaction times (RT), according to Hick-Hyman Law, for one through eight choice-RT situations, based on a simple RT of 200 ms (Magill, 2010)

MATERIALS/METHODS

Participants

Use the participants in the class.

Task and Apparatus

- Stopwatch
- Results sheet

Procedure

Break into groups of six or seven (no less than six). One person will be the recorder with the answer sheet and stopwatch. Every other group member is to face the same direction and at arm's length distance apart. The recorder will receive the randomized order of the conditions and foreperiods. Decide how long each foreperiod will be for each trial in the first condition. You will be assigned at random a condition order from the researcher, and the entire group will know this condition order.

For the simple reaction time condition, the place to be touched will be known by every member. That spot represents where everyone will touch the next person in line. For the choice reaction time conditions, all members

Name_____
Section_____ Date_____

will know the possible places that could be touched; however, only the recorder and the first person will know where he or she will touch. Where the first person touches is the spot where everyone else will touch the next person in line. For each condition, perform two practice trials and record the data as well.

For the recorder: You have the list of randomized condition orders and the stopwatch. You will decide the length of time (1–5 seconds) of the foreperiods for each condition that you tell to the first person in line. You will let the entire group know what condition number they are on before you start the practice trials. You will also let them know when you are done with the practice trials in each condition. But do not tell the group where they will be touched or the length of the foreperiod; that is only between you and the first person in line. Inform the group when you are about to start, then wait the designated time before touching the first person in line. Make sure to coordinate the timing between your touch and the start of the stopwatch. This may take a few times. Also make sure to check at the end of the trial that everyone touched the right spot. If someone touched the wrong place, that trial does not count and must be redone. Make sure to choose a new spot to touch.

Once the first person initiates movement, the time begins. Time ends when the last person in line raises his or her hand. Record this overall time on the answer sheet.

Possible places to touch for each condition:

Condition 1: only one choice

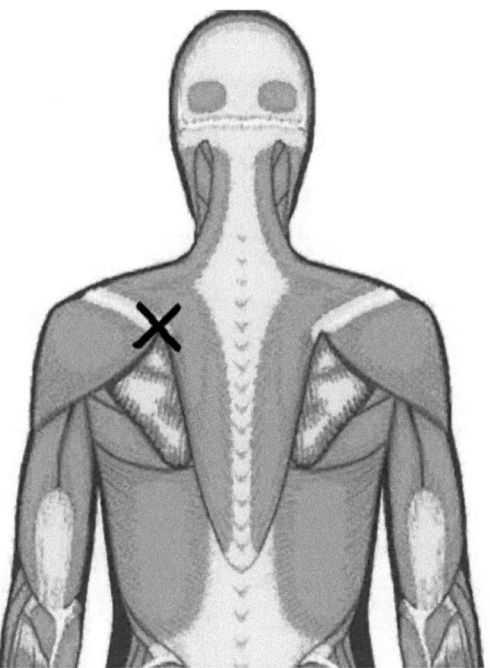

Condition 2: two possible choices

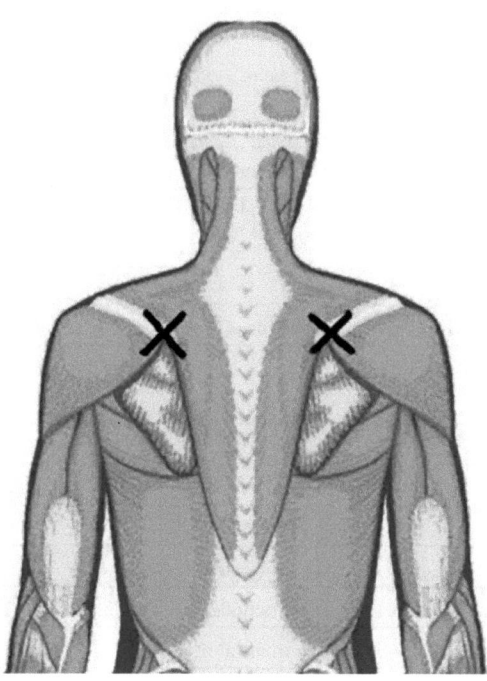

Motor Control and Motor Learning Foundations 89

Name_____
Section_____ Date_____

Condition 4: four possible choices

Condition 8: eight possible choices

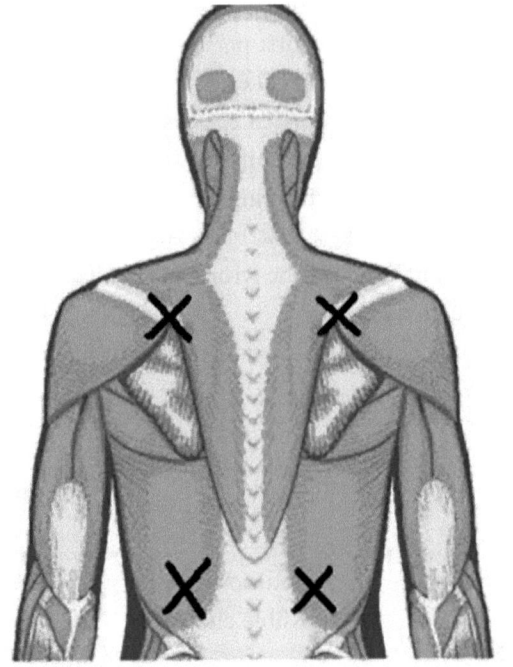 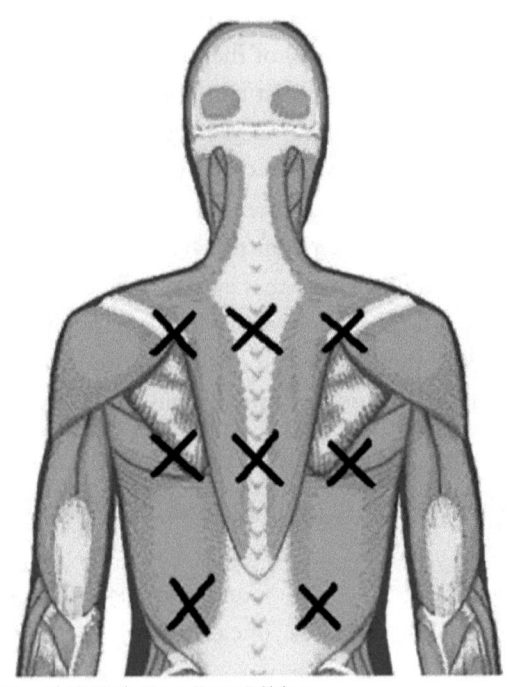

Adapted from: Richard Schmidt and Tim Lee, "Figure 3.5," Motor Control and Learning: A Behavioral Emphasis, pp. 60. Copyright © 2005 by Human Kinetics Publishers.

RESULTS

Motor Control Answer Sheet			
# in group (do not include recorder:			
Foreperiod	Condition #:	1 2 4 8	
		Overall time (in seconds)	Reaction time (in ms)
	Practice Trial		
	Practice Trial		
	Experimental 1		
	Experimental 2		
	Experimental 3		
	Experimental 4		
	Experimental 5		
	Average (Exp. only, no practice)		

Name_____

Section_____ Date_____

Foreperid	Condition #:	1 2 4 8		
		Overall time (in seconds)		Reaction time (in ms)
	Practice Trial			
	Practice Trial			
	Experimental 1			
	Experimental 2			
	Experimental 3			
	Experimental 4			
	Experimental 5			
	Average (Exp. only, no practice)			
Foreperid	**Condition #:**	1 2 4 8		
		Overall time (in seconds)		Reaction time (in ms)
	Practice Trial			
	Practice Trial			
	Experimental 1			
	Experimental 2			
	Experimental 3			
	Experimental 4			
	Experimental 5			
	Average (Exp. only, no practice)			
Foreperid	**Condition #:**	1 2 4 8		
		Overall time (in seconds)		Reaction time (in ms)
	Practice Trial			
	Practice Trial			
	Experimental 1			
	Experimental 2			
	Experimental 3			
	Experimental 4			
	Experimental 5			
	Average (Exp. only, no practice)			

Name_____
Section_____ Date_____

For estimated reaction time measures, the formula is:
RT = overall time / (# of members – 2) * **Do not include the recorder in the number of group members.**

<u>Hick-Hyman Law Graph</u>:

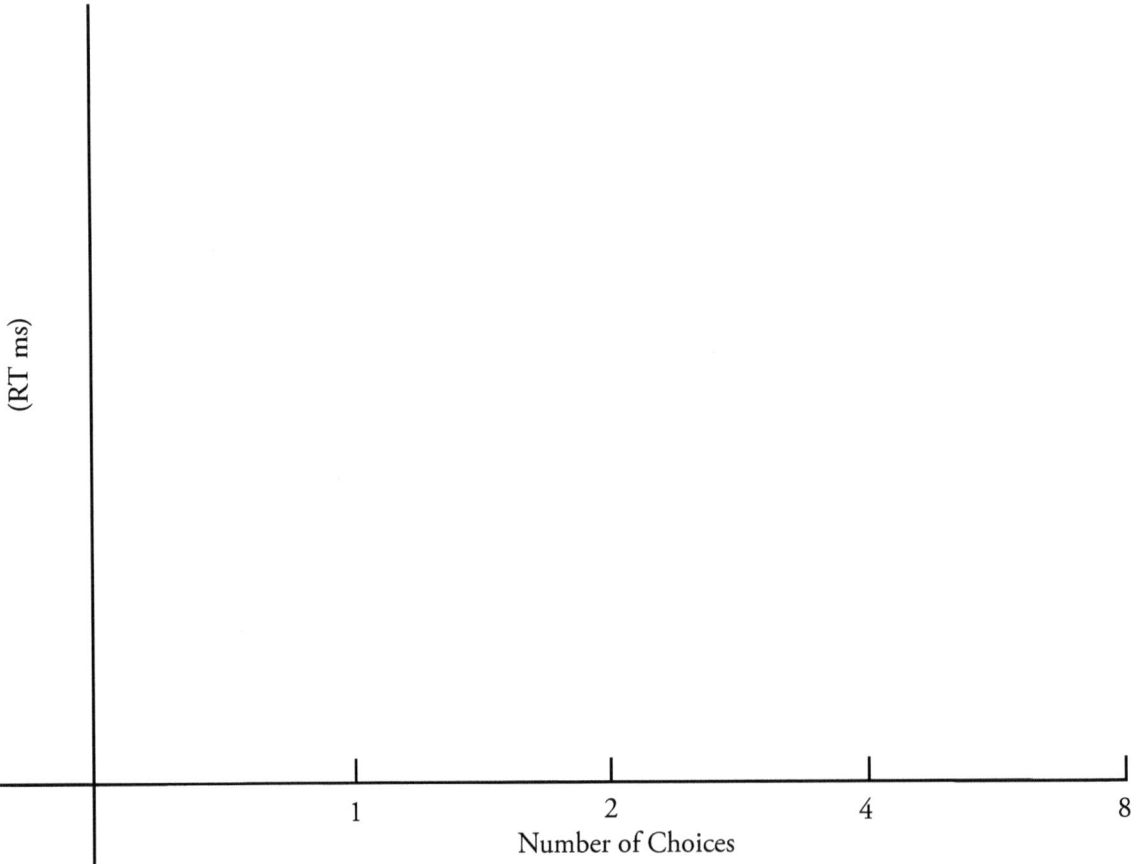

Name_____
Section_____ Date_____

DISCUSSION QUESTIONS

1. Discuss how we can use reaction time (RT) as an index of the preparation required to perform a motor skill.

2. Discuss how Hick-Hyman Law is relevant to helping understand the characteristics of factors that influence motor control preparation.

3. What is the cost-benefit trade-off involved in biasing the preparation of an action in the expectation of making one of several possible responses? Give a motor skill performance example illustrating this trade-off.

4. Describe two performer characteristics that can influence preparation. Discuss how these characteristics can influence preparation.

5. Select a motor skill and describe two motor control features of that skill that a person prepares prior to the initiation of performance of the skill.

6. What is the purpose of the foreperiods? Explain how this time affects reaction time.

REFERENCES

Hick, W. E. (1952). On the rate of gain of information. *Quarterly Journal of Experimental psychology, 4,* 11–26.

Hyman, R. (1953). Stimulus information as a determinant of reaction time. *Journal of Experimental Psychology, 43,* 188–196.

Magill, R. A. (2010). *Motor learning and control: Concepts and applications* (9th ed.). New York: McGraw Hill.

Mitchell, M. (2013). *Introduction to kinesiology: The science of human physical activity.* San Diego, CA: Cognella Academic Publishing.

Schmidt, R. A. and Lee, T. D. (2011). *Motor control and learning: A behavioral emphasis* (4th ed.). Champaign, IL: Human Kinetics.

Name_____

Section_____ Date_____

Name_____
Section_____ Date_____

CHAPTER LAB #6.2

AUGMENTED FEEDBACK: PERFORMANCE AND LEARNING

Source: http://commons.wikimedia.org/wiki/File:Darts_in_a_dartboard.jpg. Copyright in the Public Domain.

PURPOSE

In Chapter 6 in the textbook (Mitchell, 2013), we discussed the importance of different types of feedback in the performance and learning of motor skills. The aim of this activity is to highlight the important distinction between *performance* and *learning* and to show how augmented feedback influences performance and learning. A secondary aim is to practice utilizing the error formulas for AE, CE, and VE introduced in one of the activities in Chapter 1.

INTRODUCTION

Two types of feedback are typically distinguished in motor control and learning (Anderson, Magill, Sekiya, and Ryan, 2005; Christina and Shea, 1988, 1993; Wallace and Hagler, 1979; Wulf and Shea, 2002). The first is *intrinsic feedback*, which refers to information about performance that is naturally available in a task. For example, when you shoot a basketball you can see whether you made the shot or missed. The second type of feedback is *extrinsic (or augmented) feedback*, which refers to information about performance that comes from an outside source. Extrinsic feedback is further divided into *knowledge of results (KR)* and *knowledge of performance (KP)*. Knowledge of results is information on the outcome of a performance. For example, if you attempted to swim 200 m with a particular split time between the first and second hundreds, you would need a coach to provide you with knowledge of results about how successful you were in achieving your goal. Knowledge of performance is information about the movement you used to accomplish a task. A coach might tell you that you were finishing the pull phase of your stroke too early during your 200 m swim.

In this lab you will analyze performance and learning in an aiming task with two different types of KR. This lab is designed to highlight two important concepts in motor learning, and to provide you with practice in applying the error formulas for AE, CE, and VE described above. The first important concept in this lab is the distinction between performance and learning. Performance refers to temporary changes in behavior whereas learning refers to relatively permanent changes in behavior. We will assess the difference between the two by comparing practice with KR versus retention without KR. The second important concept is that learners can become dependent on KR when it is provided too frequently or in a manner that is too easy to use. We will test this notion by comparing a group that receives KR directly after each practice attempt relative to a group that receives KR after a delay of two trials.

Name_____
Section_____ Date_____

MATERIALS/METHOD

Participants

Everyone in the class can participate in this activity.

Task and Apparatus

Any aiming task can be used for this lab, such as throwing darts at a target, tossing a beanbag toward a target, putting a golf ball, and so on. The key is that the performer is blindfolded and reliant on an experimenter to provide KR about where each attempt landed relative to the target. We have chosen to use aiming blindfolded at a target on a piece of paper. The target, located at the end of this lab, can be taped to any flat surface including the floor, a tabletop, or a chair with a desk arm. The performer takes a pencil and places it on the start location of the target. The goal is to make a smooth, continuous movement to hit the target.

Procedure

Data will be collected in pairs. One person in the pair will be the experimenter and the other person will be the learner. The pairs will switch roles after Phase I in the experiment and then switch back for Phase II. Phase I involves 30 practice trials with KR, and Phase II involves 10 retention trials without KR. The retention trials should be done after a 10-minute break, so the first learner can perform the retention test after the second learner has finished his or her practice trials. The second learner can perform the retention trials approximately 5 minutes after the first learner has finished their retention trials.

Pairs will flip a coin to see who will be assigned to the *immediate* or the *delayed* KR group. The experimenter will record the error (with the + or – sign) after each trial. The immediate group will receive their KR immediately after each practice trial; however, the delayed group will receive their KR after a delay of two trials (e.g., they will get KR for Trial 1 after Trial 3, KR for Trial 2 after Trial 4, and so on. Neither group will receive KR in retention.

Data Analysis

Record your data in the table below. Calculate AE, CE, and VE for each block and insert the values into the appropriate cells of the table. Neatly draw three separate line graphs, one each for AE, CE, and VE, on the following sheet of paper.

Name_____
Section_____ Date_____

GROUP _____

TRIALS	BLOCK 1	BLOCK 2	BLOCK 3	10-min break	RETENTION
TRIAL 1					
TRIAL 2					
TRIAL 3					
TRIAL 4					
TRIAL 5					
TRIAL 6					
TRIAL 7					
TRIAL 8					
TRIAL 9					
TRIAL 10					
$\sum |e|$					
$\sum (e)$					
AE					
CE					
VE					

RESULTS

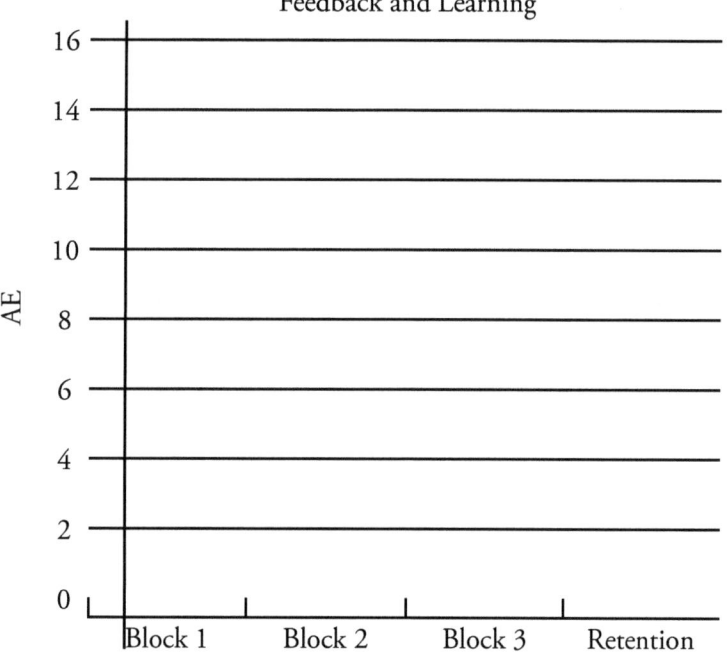

Feedback and Learning

Motor Control and Motor Learning Foundations

Name_____
Section_____ Date_____

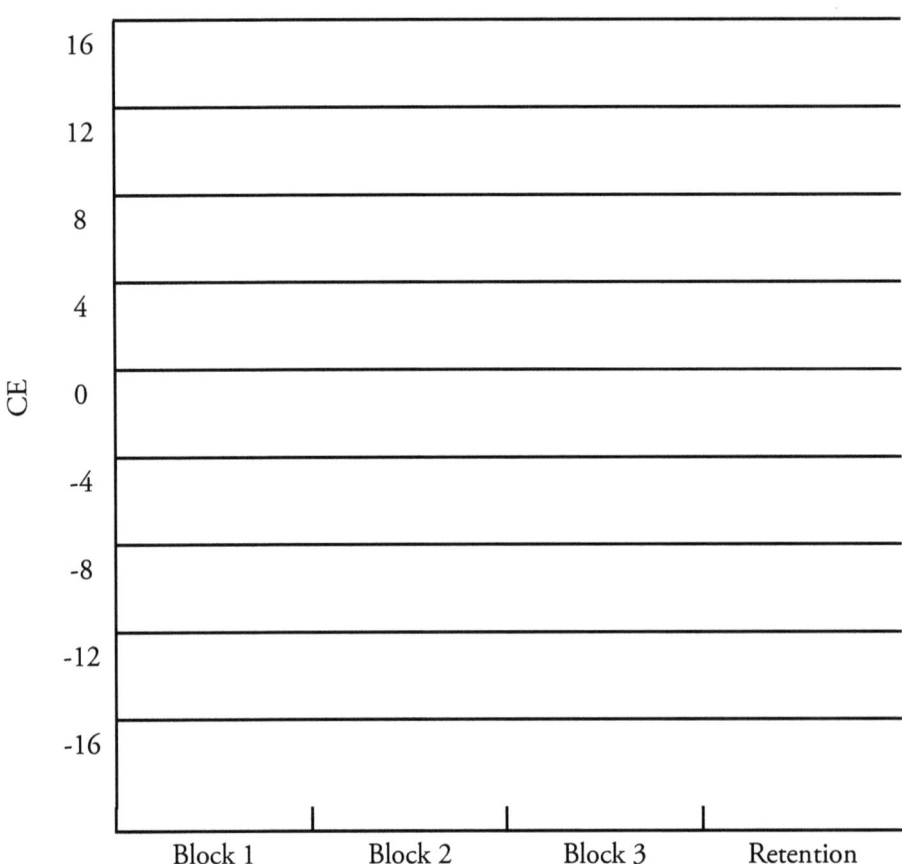

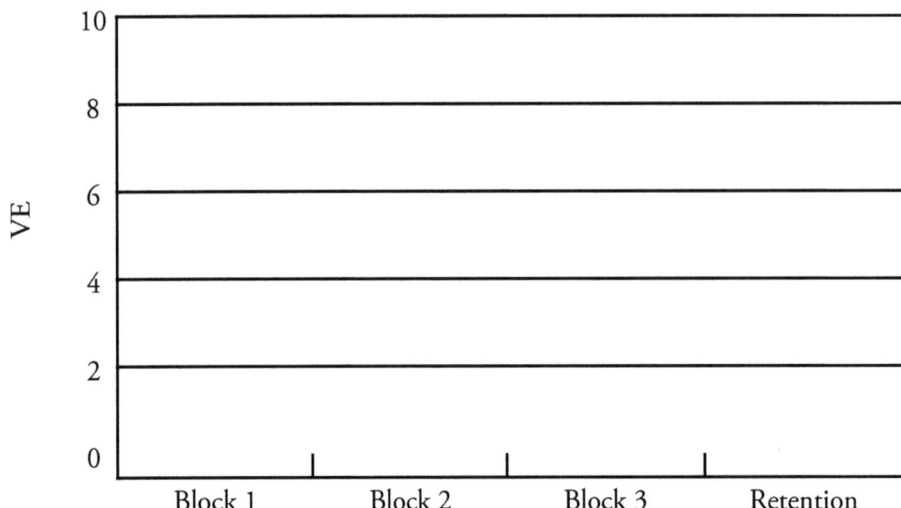

Name_____
Section_____ Date_____

DISCUSSION QUESTIONS

1. Use your data to describe the magnitude, bias, and consistency of your errors during practice.

2. Was your practice performance a good indicator of how well you learned the task (i.e., how well you performed in retention)? Why?

3. Why might we want to use more than one type of error to characterize performance and learning?

4. Why are changes in performance not a good indicator of learning?

5. Aside from using retention tests, how else could you determine how well someone has learned a task?

6. With respect to augmented feedback, what is the guidance hypothesis?

7. What are some ways to minimize the negative effects of guidance when providing augmented feedback?

8. Are the principles that apply to the use of KR with simple skills likely to apply to the use of KP with more complex skills?

Name_____
Section_____ Date_____

REFERENCES

Anderson, D. I., Magill, R. A., Sekiya, H., and Ryan, G. (2005). Support for an explanation of the guidance effect in motor skill learning. *Journal of Motor Behavior, 37,* 231–238.

Christina, R. W., and Shea, J. B. (1988). The limitations of generalization based on restricted information. *Research Quarterly for Exercise and Sport, 59,* 291–297.

Christina, R. W., and Shea, J. B. (1993). More on assessing the retention of motor learning based on restricted information. *Research Quarterly for Exercise and Sport, 64,* 217–222.

Mitchell, M. (2013). *Introduction to kinesiology: The science of human physical activity.* San Diego, CA: Cognella Academic Publishing.

Wallace, S. A., and Hagler, R. W. (1979). Knowledge of performance and the learning of a closed motor skill. *Research Quarterly, 50,* 265–271.

Wulf, G, and Shea, C. H. (2002). Principles derived from the study of simple skills do not generalize to complex skill learning. *Psychonomic Bulletin and Review, 9,* 185–211.

Name_____
Section_____ Date_____

+21	
+20	
+19	
+18	
+17	
+16	
+15	
+14	
+13	
+12	
+11	
+10	
+9	
+8	
+7	
+6	
+5	
+4	
+3	
+2	
+1	
-1	
-2	
-3	
-4	
-5	
-6	
-7	
-8	
-9	
-10	
-11	
-12	
-13	
-14	
-15	
-16	
-17	
-18	
-19	
-20	
-21	
-20	
-21	

Name_____

Section_____ Date_____

Chapter Seven
Psychological Foundations

CHAPTER ACTIVITY #7

MOTIVATION AND PHYSICAL ACTIVITY

Source: http://commons.wikimedia.org/wiki/File:KBryant8.jpg. Copyright in the Public Domain.

PURPOSE
The purpose of this activity is to review material presented in Chapter 7 of the textbook (Mitchell, 2013) and to learn more about the motivations related to participation in physical activity.

INTRODUCTION
In Chapter 7, we examined a number of concepts and issues related to psychological factors involved in physical activity. The major topics discussed were motivation, arousal, performance enhancement, and exercise behavior. Motivation is particularly important in understanding why people participate in physical activity, and why they

Name_____
Section_____ Date_____

Name_____

Section_____ Date_____

adopt and discontinue an exercise program. After answering the following chapter summary questions, it's hoped that this activity will allow you to acquire better insight into the motivations affecting participation in physical activity.

CHAPTER SUMMARY QUESTIONS

1. Identify three physical activities that you enjoy. Describe the motives for why you participate in these activities.

   Activity #1 _____

   Motives: _____

   _____

   _____

   Activity #2 _____

   Motives: _____

   _____

   _____

   Activity #3 _____

   Motives: _____

   _____

   _____

2. For activity #1, what is your motivational orientation? Why do you have this type of orientation?

   Motivational orientation: _____

   Reason for having this orientation: _____

   _____

   _____

Name_____
Section_____ Date_____

3. What type of motivational climate existed in your family that may have influenced the type of physical activities you now participate in?

   _____

   _____

   _____

4. Briefly compare the optimal level of arousal necessary for lifting weights to the optimal level of arousal for putting in golf. Explain your answer:

   _____

   _____

   _____

5. Pick one type of physical activity that you are trying to improve upon. Provide three types of goals that you could use to help you improve.

   Physical activity: _____

   Three goals: _____

   _____

   _____

6. Let us say you have a family member or a friend who is having difficulty adhering to exercise and is thinking about dropping their membership to the local health club. Using what you have learned in sport and exercise psychology, what three suggestions can you make to help this person adhere to an exercise program?

   _____

   _____

   _____

Name_____
Section_____ Date_____

CHAPTER ACTIVITY

Equipment and Materials Needed: Materials to Take Notes (paper and pencil, laptop computer, etc.)

The purpose of this activity is to determine whether the sex of an individual may affect participation in physical activity. This activity will require you to interview two people, one male and one female. Make sure the two individuals are of approximately the same age. For example, you could choose to interview a male and female in their late childhood, early adulthood, or late adulthood. You should arrange to conduct the interviews at different times but within a few days of each other. The major interview objectives involve determining what type of physical activities the individuals now participate in, whether the activities have changed over the years, what types of motives were involved in each activity and what type of motivational climate the individuals grew up in (see pages 208–211 in the textbook), and what types of positive and negative experiences they had in each activity.

PROCEDURES

1. After choosing the two individuals, arrange a meeting time for each. Inform them that the interview is part of an assignment associated with this class, and that their names will remain anonymous. Also, tell them that the interview will be conducted within a one-hour period. Arrange to meet them at a convenient location that is quiet and relatively free from distractions.
2. Based upon the follow categories, develop a list of questions to ask each person:

   - Previous physical activities (exercise, sports, recreational, hobbies, etc.)
   - Current physical activities
   - Motives for participation in each activity, including whether they are intrinsically or extrinsically motivated
   - Motivational climate growing up (intrinsic or extrinsic)
   - Positive and negative experiences in each activity

3. Have the instructor in your course look over your questions and make any suggested changes.
4. At the interviews, make sure you have plenty of paper and writing materials. Use a laptop computer or notepad if you feel this would work better for you. Ask permission if you plan to use some type of recording equipment to record the interview.
5. After each interview, briefly edit and summarize the answers.
6. Prepare a report to be turned in to your instructor.
7. Make sure that your report discusses any differences between the male and female participants.
8. Are you able to conclude that one's sex is a contributing factor to their participation in physical activity?

REFERENCE

Mitchell, M. (2013). *Introduction to kinesiology: The science of human physical activity.* San Diego, CA: Cognella Academic Publishing.

Name_____

Section_____ Date_____

Name_____
Section_____ Date_____

CHAPTER LAB #7.1

GOAL SETTING, FEEDBACK, AND PERFORMANCE

- This lab is a modification of a lab originally designed by Nick O'Dwyer and Roger Adams at the University of Sydney, Faculty of Health Sciences.

PURPOSE

The aim of this activity is to examine how goal setting and augmented feedback influence performance on a muscular endurance task.

INTRODUCTION

In an earlier lab you learned that augmented feedback can play an important role in guiding motor learning if it is used judiciously. Augmented feedback also plays an important motivational role, particularly when used with goal setting. Both variables can independently help motivate people to achieve their best performance or to acquire or re-acquire a new skill (Bickers, 1993; Locke and Latham, 2002; Locke, Shaw, Saari and Latham, 2000; McNair, Depledge, Brettkelly and Stanley, 1996; Moffatt, Chitwood and Biggerstaff, 1994; Tubbs, 1986; Weinberg, Bruya, Garland, and Jackson, 1990). When used together, however, their effect is more potent. In Chapter 7 of the textbook (Mitchell, 2013), we discussed the importance of goal setting. Goal setting is superior to simply asking someone to do their best, because it establishes a specific target toward which the person can strive. The target then helps to direct the person's effort. Feedback tells the person how close they are to achieving their goal, and therefore specifies how much more effort is needed for success.

MATERIALS/METHOD

Participants

Each person in the class will participate in the experiment, unless he or she has a musculoskeletal problem that prevents attempting the wall squat.

Name_____
Section_____ Date_____

Task and Apparatus

The wall-squat endurance task will be used to examine the effects of goal setting and augmented feedback on performance, though any task that assesses muscular endurance could potentially be used. The goal is to maintain the squat for as long as possible. The participant squats as shown in the figure, with the back flat against the wall, thighs horizontal, shanks vertical, and feet flat on the floor. The arms should be folded across the chest. Timing is started as soon as the participant assumes the position and is stopped when the participant can no longer maintain it. ***Stopwatches*** are used to measure endurance times and a ***laptop or electronic device*** with a spreadsheet application is needed to record data. The stopwatch on a cell phone provides an ideal timing device.

Procedure

The experiment has two phases. Phase 1 will be performed today, and Phase 2 will be performed next week. The measurement group is responsible for instructing each participant on what to do, timing the squat, and recording each participant's data. The measurement group should be briefed on the procedure before the class is allowed into the testing area. The class should be brought into the testing area in small groups so that there is only one participant for each measurement person at any given time.

PHASE I

In **Phase 1**, the procedure is the same for all three experimental groups, as follows:

Control/nonspecific goal—Participants are instructed simply to "maintain the squat for as long as possible." They are given no encouragement of any kind and all other students must refrain from conversation during the performance. One student directs the participant and records the participant's name and time. **Do not inform the participants of their times!** It is preferable if each group of performers does not start at exactly the same time so that they don't compete with each other.

Once every participant has been tested, the measurement group should input the participant names and times into a spreadsheet. The times should be rank-ordered and the participants divided into three even groups (i.e., the average squat time should be equivalent for each group). One way to ensure even groups is to simply assign a number from 1 to 3 in a column next to the rank-ordered times. The numbers should be assigned in the following sequence: 1, 2, 3, 3, 2, 1, 1, 2, 3, 3, 2, 1, 1, 2, 3, 3, 2, 1.... until you get to the last row in the data set. If the spreadsheet is then sorted based on the column with the 1s, 2s and 3s, the participants in each group will be blocked together and the mean endurance times for each of the three groups should be approximately equal.

Participants in Group 1 are the ***Control/nonspecific*** *goal* group.
Participants in Group 2 are the ***Specific goal*** group.
Participants in Group 3 are the ***Specific goal/performance feedback*** group.

For participants in Groups 2 and 3, **create another column in your spreadsheet and multiply their original squat time by 1.5. This is the goal they will strive to achieve next week.**

PHASE II

In this activity we complete Phase II of our goal-setting, feedback, and performance lab. The groups will be tested sequentially to avoid spillover effects from one experimental manipulation to another. All participants should wait outside until called by name by the measurement group.

Name_____

Section_____ Date_____

Group 1 (control/nonspecific goal)—The procedure for Group 1 is exactly the same as that used last week. The participants are instructed simply to "maintain the squat for as long as possible" and are given no encouragement or feedback of any kind. Again, one student directs the participant and records their name and time.

Group 2 (specific goal)—In this group the participants are instructed to maintain the squat for a specific duration. This duration will be half as long again as they achieved in the first trial (trial 1 duration x 1.5). The participants are instructed to "maintain the squat for at least [goal duration] seconds." Thus, this will be a specific hard goal. Each participant will be given their own individualized goal based on their first trial performance. Participants are not given any feedback about how much longer they have to go before they attain their goal nor are they told when they have attained their goal. In addition, they are given no encouragement of any kind.

Group 3 (specific goal/feedback)—In this group the participants are also instructed to maintain the squat for at least 1.5 times the duration of their Trial 1 performance, and they are given performance feedback when they request it. The experimenter can tell them how much time has already elapsed since the start of the trial but again they are given no encouragement of any kind. Note that the participants do not have to stop when the goal is achieved but may continue for longer if possible.

Data Analysis

The measurement group should record both trials in their spreadsheet and calculate the percentage change from Trial 1 to Trial 2.

RESULTS

Mean duration and % change from trial 1 to trial 2

	Trial 1 (seconds)	Trial 2 (seconds)	% Change $\frac{(Trial\ 2 - Trial\ 1)}{Trial\ 1}$
Control			
Specific Goal			
Specific Goal/Feedback			

The measurement group should use this table to organize their data. The percentage change data should be plotted on a bar graph that includes all three groups.

Name_____
Section_____ Date_____

DISCUSSION

Goal Setting

While goal setting can be a very effective motivator, the setting of appropriate goals is crucial. Appropriately set and achieved goals can increase the likelihood of a person remaining involved in an activity or rehabilitation program. Tubbs (1986) analyzed and summarized the research literature as revealing four goal-setting characteristics that influence performance of skills in a positive manner:

1. Difficult goals lead to better performance than easy goals.

2. Specific goals lead to better performance than do-your-best goals or no goals.

3. Goal setting plus performance feedback (information about performance specifically in relation to the set goal) is better than goal setting alone.

4. Participant involvement in goal setting may lead to better performance than goals assigned without participant involvement (participant involvement does not mean that the participants set the goal but that they are involved in the goal-setting process with the teacher).

Setting specific, challenging goals is best for motivating performance (Locke et al., 2000). Although questions remain about *why* goal setting is effective, the important point here is that goal setting *is* effective.

Name_____
Section_____ Date_____

DISCUSSION QUESTIONS

1. Provide a descriptive overview of the findings from the entire class.

2. Was there any change in the performance of the control group on this motor task? What might account for an improvement in the control group?

3. Why are "do your best" goals usually not as effective as specific, objective goals in enhancing the performance of a motor task? Were the specific goals set here difficult enough?

4. How might we have made the goal-setting process more effective?

5. How do you think performance feedback works to enhance task performance?

6. How do our results fit with the features of Locke and Latham's (2002) theory of goal setting? Could we have realized even better results by using unrealistically high goals?

7. How might verbal encouragement have influenced the results?

Name_____

Section_____ Date_____

REFERENCES

Bickers, M. J. (1993). Does verbal encouragement work? The effect of verbal encouragement on a muscular endurance task. *Clinical Rehabilitation, 7*, 196–200.

Locke, E. A., and Latham, G. P. (2002). Building a practically useful theory of goal setting and task motivation. *American Psychologist, 57*, 705–717.

Locke, E. A., Shaw, K. N., Saari, L. M., and Latham, G. P. (2000). Goal-setting and task performance: 1969–1980. *Psychological Bulletin, 90*, 125–152.

McNair, P. J., Depledge, J., Brettkelly, M., and Stanley, S. N. (1996). Verbal encouragement: Effects on maximum effort voluntary muscle action. *British Journal of Sports Medicine, 30*, 243–245.

Mitchell, M. (2013). *Introduction to kinesiology: The science of human physical activity.* San Diego, CA: Cognella Academic Publishing.

Moffatt, R. J., Chitwood, L. F., and Biggerstaff, K. D. (1994). The influence of verbal encouragement during assessment of maximal oxygen uptake. *The Journal of Sports Medicine and Physical Fitness, 34*, 45–49.

Tubbs, M. E. (1986). Goal setting: a meta-analytic examination of the empirical evidence. *Journal of Applied Psychology, 71*: 474–483.

Weinberg, R., Bruya, L., Garland, H., and Jackson, A. (1990). Effect of goal difficulty and positive reinforcement on endurance performance. *Journal of Sport and Exercise Psychology, 12*, 144–156.

Name_____
Section_____ Date_____

CHAPTER LAB #7.2

OBSERVATIONAL LEARNING AND MENTAL REHEARSAL

Photos by authors.

PURPOSE

To examine the influence of watching a "learning model" and using mental rehearsal on motor skill acquisition and retention.

INTRODUCTION

A common way to introduce learners to a new skill is to have them watch another person perform the skill. The other person is typically someone who has a reasonable degree of proficiency in the skill. Considerable evidence suggests that watching a demonstration from a skilled model can facilitate the rate at which new skills are acquired and the degree to which they are retained after practice (McCullagh and Weiss, 2001). But research has also shown that learners can profit from watching other learners as they practice a new skill, particularly when they see or hear the feedback given to the other learner (Hebert and Landin, 1994). Observing another person perform a skill is particularly helpful when a new pattern of coordination has to be learned (Ashford, Bennett, and Davids, 2006).

In Chapter 7 of the textbook (Mitchell, 2013), it was pointed out that mental practice is another effective way to enhance performance and learning, particularly when it is used as part of a more general practice routine (Singer, 1988; Lidor, Tennant, and Singer, 1996). Cognitive rehearsal is also thought to be an important process in the retention of information conveyed by a demonstration (Bandura, 1986). Consequently, combining observational learning and mental rehearsal may be more effective than using either one independently.

In this lab you will analyze the effects of watching a *learning model* and *mental rehearsal* on the acquisition and retention of a novel skill.

MATERIALS/METHOD

Participants

Everyone in the class should participate in this experiment. The class will be paired up for data collection.

Name_____

Section_____ Date_____

Task and Apparatus

The task is the stick test of coordination. The goal of the task is to flip a pencil 180 degrees and catch it with two other pencils. Any type of stick will work (e.g., chopsticks, coffee stirring sticks, tongue depressors, as long as everyone uses the same type of stick). The task can be performed in either the seated or standing position, though the position should be standardized for everyone in the experiment. The pencil is placed initially on top of the other two pencils so that the three pencils form an H. The support pencils are held in each hand so that they point away from the body in the sagittal plane, and the other pencil sits on top of them such that its long axis is oriented in the frontal plane. The pencil can be flipped from left to right or from right to left. The learner scores two points if the pencil is caught cleanly, one point if the pencil touches any part of the hand before being rebalanced, and no points if it is dropped.

Procedure

The class is divided into pairs and the pairs flip a coin to determine who will perform the task first, the *learning model*, and who will observe and record data first and perform the task second, the *observer/imager*. Each person will perform 10 blocks of 10 trials (100 trials total). The *learning model* goes first, and the observer watches closely and records the score for each of the *learning model's* trials. A one-minute break is given between blocks of trials.

During each one-minute break, the *observer/imager* should imagine performing the task from an *internal* (first-person) perspective. That person (observer/imager) should sit comfortably with eyes closed and imagine holding the support pencils and flipping the pencil that rests on them. It is important to imagine all the sensations that would be experienced when performing the task, including the "feel" of the pencils and the body movements, the sight of the pencils and limbs, and the sound of the pencils being caught. It is equally important not to perform any overt movements.

After the *learning model* has completed all of his or her trials, the *observer/imager* then completes his or her trials, with the *learning model* then becoming the *observer* and the recorder. When the observer has finished his or her trials, the original *learning model* completes a retention test consisting of 2 blocks of 10 trials. About 10 minutes later (to ensure the retention interval is similar for both groups) the original *observer/imager* completes the retention test.

Name_____
Section_____ Date_____

Data Analysis

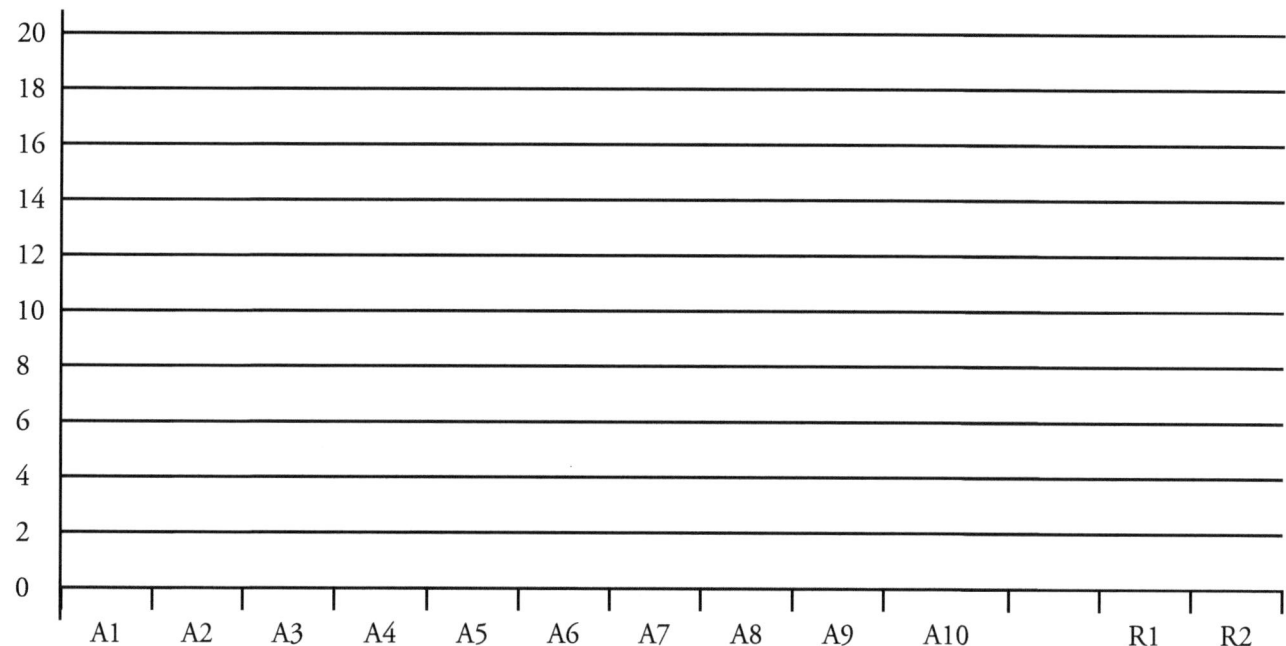
Observational Learning

Name_____
Section_____ Date_____

RESULTS

Use the line graph below to plot your individual data. Combine all participants' data and compare the *learning models'* group to the *observers/imagers'* group in acquisition and retention.

Record the scores for each block of 10 trials in the table below.

Group _____

Block	Trial 1	Trial 2	Trial 3	Trial 4	Trial 5	Trial 6	Trial 7	Trial 8	Trial 9	Trial 10	Total
A1											
A2											
A3											
A4											
A5											
A6											
A7											
A8											
A9											
A10											
Retention Interval (approx 20 min)											
R1											
R2											

Name_____
Section_____ Date_____

DISCUSSION QUESTIONS

1. Describe the effect of observational learning and mental rehearsal on acquisition performance.

2. Describe the effect of observational learning and mental rehearsal on retention performance.

3. Why might the effects on retention performance have been different from the effects on acquisition performance (assuming this was the case)?

4. Irrespective of your results, why do you think that observing a learning model might facilitate motor learning?

5. What other variables (e.g., characteristics of the model, characteristics of the learner, characteristics of the to-be-learned skill) might influence the effectiveness of demonstrations on motor learning?

6. Are there potential negative effects of using modeling/demonstrations to facilitate motor learning?

7. How could you modify the experiment to test the independent effects of demonstration and mental rehearsal on performance and learning?

8. How might individual differences in the ability to imagine and control visual and kinesthetic imagery have influenced the effectiveness of mental rehearsal? (see Goss, Hall, Buckolz and Fishburne, 1986.)

9. Can you think of other methodological problems that make studying mental rehearsal difficult?

10. When might mental practice be a good alternative or supplement to physical practice?

Name_____

Section_____ Date_____

REFERENCES

Ashford, D., Bennett, S. J., and Davids, K. (2006). Observational modeling effects for movement dynamics and movement outcome measures across differing task constraints: A meta-analysis. *Journal of Motor Behavior, 38,* 185–205.

Bandura, A. (1986). *Social foundations of thought and action: A social-cognitive theory.* Englewood Cliffs, NJ: Prentice Hall.

Feltz, D. L., and Landers, D. M. (1983). The effects of mental practice on motor skill learning and performance: A meta-analysis. *Journal of Sport Psychology, 5,* 25–57.

Goss, S., Hall, C., Buckolz, E., and Fishburne, G. (1986). Imagery ability and the acquisition and retention of motor skills. *Memory and Cognition, 14,* 469–477.

Hebert, E. P., and Landin, D. (1994). Effects of a learning model and augmented feedback on tennis skill acquisition. *Research Quarterly for Exercise and Sport, 65,* 250–257.

Hird, J. S., Landers, D. M., Thomas, J. R., and Horan, J. J. (1991). Physical practice is superior to mental practice in enhancing cognitive and motor task performance. *Journal of Sport and Exercise Psychology, 8,* 282–293.

Lidor R., Tennant, K. L., and Singer, R. N. (1996). The generalizability effect of three learning strategies across motor task performances. *International Journal of Sport Psychology, 27,* 23–36.

McCullagh, P., and Weiss, M.R. (2001). Modeling: Considerations for motor skill performance and psychological responses. In R.N. Singer, H.A. Hausenblas, and C.M. Janelle (Eds.), *Handbook of sport psychology* (2nd ed.) (pp. 205–238). New York: Wiley.

Mitchell, M. (2013). *Introduction to kinesiology: The science of human physical activity.* San Diego, CA: Cognella Academic Publishing.

Singer, R. N. (1988). Strategies and metastrategies in learning and performing self-paced athletic skills. *The Sport Psychologist, 2,* 49–68.

Chapter Eight
Developmental Foundations

CHAPTER ACTIVITY #8

COMPARING MALE AND FEMALE WORLD RECORDS

PURPOSE
The purpose of this activity is to review material presented in Chapter 8 of the textbook (Mitchell, 2013) and to explore some predictions of an interesting study by Whipp and Ward (1992) regarding male and female world record performance.

INTRODUCTION
In Chapter 8 we reviewed several concepts related to the human development across the life span. One of the important concepts was the examination of differences in performance between male and females. We argued that these differences are apparent during childhood and are evident in adulthood. But what about differences between males and females who are elite or world-class athletes? This activity allows you to explore the differences in performance between male and female elite athletes and to better understand why these differences may exist.

Name_____

Section_____ Date_____

Name_____
Section_____ Date_____

CHAPTER SUMMARY QUESTIONS

1. If a longitudinal study on sixth-graders is not practical to do over a three-year period, what other kind of research design could help? What type of groups of participants would you need to observe?

   Design type: _____

   Groups needed: _____

   _____

2. Based on your knowledge of the development of the skeletal system, name one type of physical activity that should be discouraged for a growing child. Name three activities that would be encouraged.

   Discouraged activity: _____

   Encouraged activities: _____

   _____

   _____

3. Provide one example of each type of reflex (primitive, postural, and locomotor). What is the stimulus for each reflex and what is the response?

   Primitive:_____

   Stimulus _____

   Response _____

   Postural:_____

   Stimulus _____

   Response _____

   Locomotor:_____

   Stimulus _____

   Response _____

Name_____
Section_____ Date_____

4. Name one important decline that occurs as a function of aging in the skeletal, muscular, nervous system, and cardiovascular/pulmonary system.

   Skeletal system decline: _____

   Muscular system decline: _____

   Nervous system decline: _____

   Cardiovascular/pulmonary system decline: _____

   _____

5. What are the three major elements of physical health? Provide an example of each:

   Element #1 _____

   Example _____

   Element #2 _____

   Example _____

   Element #3 _____

   Example _____

CHAPTER ACTIVITY

Equipment and Materials Needed: Access to a Library and the Internet

This activity will require you to do some library work. In the Whipp and Ward (1992) study reviewed in your textbook (pages 253–254), the authors reviewed world-class records set by men and women in running events. Review the discussion in the text, then go to your university library and find the actual paper:

Whipp, B. J. and Ward, S.A. (1992). Will women soon outrun men? *Nature, 255*, 25.

After reading this paper, read the following article:

Seiler, S. and Sailer, S. (1997). The gender gap: Elite women are falling further behind. *Sportscience News,* May–June.

This article can be viewed online:
http://www.sportsci.org/news/news9705/gengap.html

Name_____
Section_____ Date_____

After you read both the Whipp and Ward (1992) and Seiler and Sailer (1997) articles, please answer the following questions:

DISCUSSION QUESTIONS

1. How have men's running records changed since records have been kept?

2. How have women's records changed?

3. What predictions were made by Whipp and Ward (1992) about men's and women's performance in the future?

4. According to the Seiler and Sailer (1997) article, have these predictions been supported?

5. What are some of the physiological differences between males and females that could account for performance differences?

6. What are some sociocultural factors that may influence women's elite running performances relative to men's?

7. Have performance enhancing drugs affected male and female running performances? If so, in what way?

REFERENCES

Seiler, S. and Sailer, S. (1997). The gender gap: Elite women are falling further behind. *Sportscience News*, May–June. (http://www.sportsci.org/news/news9705/gengap.html)

Mitchell, M. (2013). *Introduction to kinesiology: The science of human physical activity.* San Diego, CA: Cognella Academic Publishing.

Whipp, B. J. and Ward, S.A. (1992). Will women soon outrun men? *Nature, 255,* 25.

Name_____

Section_____ Date_____

Name_____
Section_____ Date_____

CHAPTER LAB #8.1

ASSESSING FUNDAMENTAL MOTOR SKILLS

Copyright © 2010 Depositphotos/Miroslaw Dziadkowiec.

PURPOSE
The aim of this activity is to develop skill in assessing fundamental motor skills and to identify factors that might explain gender differences in the performance of fundamental motor skills.

INTRODUCTION
Fundamental motor skills (FMS) are patterns of movement typically acquired by children after the rudimentary postural, locomotor, and manipulative (prehension) skills of toddlerhood have been learned. Two broad classes of skills are acquired: those that involve locomotion, such as jumping, hopping, galloping and skipping; and those that involve the projection and interception of objects. FMS are important because they are thought to serve as the building blocks for more advanced skills that are used in sports and games (Haywood and Getchell, 2009). In addition, a broader repertoire of FMS is thought to increase the likelihood that children will engage in physical activity throughout the life span (Stodden et al., 2008).

Developmental sequences are commonly used to assess FMS competence. These sequences are qualitative descriptions of how body parts move when the child has acquired different levels of skill mastery. Sequences have been developed for the whole body (*total body sequences*) and for the individual body segments (*component sequences*). The configuration or movement of the entire body at each developmental level is described in the total body approach. The configuration or movement of individual body segments is described in the component approach, such that one segment (e.g., arms) could be at a different level from another segment (e.g., legs) at any point in time.

Overarm throwing is a skill that has been studied extensively by developmental researchers, dating back to the early work of Monica Wild (Wild, 1938). Total body sequences (e.g., Seefeldt, Reuschlein, and Vogel, 1972) and component sequences (e.g., Roberton and Halverson, 1984) have been developed for throwing. In this activity, you will assess the throwing level of your peers using a version of the total body approach developed by Seefeldt et al. (1972). In addition, you will look for gender differences in skill level and discuss potential explanations for any differences you observe (see Hardy et al., 2010; Thomas and French, 1985; Thomas and Marzke, 1992; Mitchell, 2013 [see Chapter 8, pages 252–253]).

Name_____
Section_____ Date_____

MATERIALS/METHOD

Participants

Every member of the class should participate in this activity. The activity is best done in groups of four students.

Task and Apparatus

The performer's task is to throw a tennis ball six times toward a target approximately 10 m away. The throw will be performed three times with the **dominant arm** and three times with the **nondominant arm**. The goal is to throw the ball with maximum velocity. The assessors' task is to observe the three throws with each arm from three different vantage points (behind, in front of, and to the side of the thrower), and determine what level of competence the thrower possesses. The levels are described in the table below.

Procedure

Students should familiarize themselves with the developmental sequence for throwing (table below) before any activity begins. Prior to throwing, students should engage in a warm up that involves 5 minutes of light aerobic activity followed by 5 minutes of stretching. Each person should complete two throws at 25% of max velocity, two throws at 50% max velocity, and two throws at 75% max velocity with each arm before attempting to throw at maximum velocity. Students in each group should take turns performing and assessing the throws. When one person is throwing, the other three members of the group can station themselves at each of the viewing positions and change positions every three throws. After one person has completed all of their six throws, the assessors should average their ratings for each throw and then get together and average their ratings for that person.

Ratings should be based on the descriptions in the following table:

Developmental Level				
1	2	3	4	5
Chop	*Slingshot*	*Ipsilateral Step*	*Contralateral Step*	*Windup*
Vertical windup "chop" throw Feet stationary No spinal rotation	Horizontal windup "Slingshot throw" Block rotation Follow-through across body	High windup Ipsilateral step Little spinal rotation Follow-through across body	High windup Contralateral step Little spinal rotation Follow-through across body	Downward arc windup Contralateral step Segmented body rotation Arm-leg follow-through

Data Analysis

The ratings should be analyzed for the entire class. Calculate the mean ratings and the standard deviations of the ratings for the males and females in the class for each arm.

Name_____
Section_____ Date_____

RESULTS

Plot the class means and their standard deviations for the males and females for each arm on the following graph.

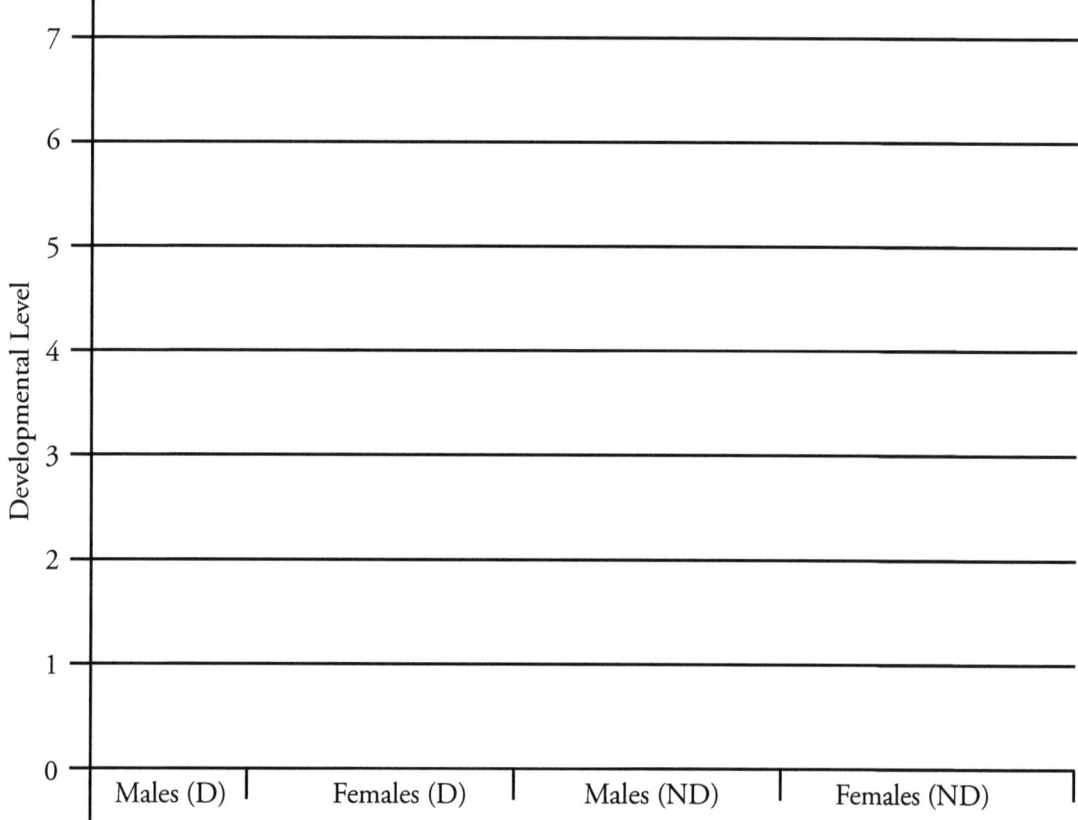

Developmental Foundations 129

Name_____
Section_____ Date_____

DISCUSSION QUESTIONS

1. Do you think that a total body approach or a component approach would be better for analyzing competence in a skill like throwing? Explain your answer.

2. How would you divide the body (or segments of the throwing movement) if you were asked to create a component sequence for throwing (i.e., determining how many components to analyze and what they would be)?

3. Compare the differences between males and females on their level of competence in throwing.

4. Throwing is a task where boys typically show advanced development relative to girls, particularly after puberty. What environmental factors (including sociocultural factors) might account for differences between boys and girls?

5. What anatomical differences might account for differences in throwing proficiency between boys and girls?

6. Do the male–female differences between the dominant and nondominant arms provide clues as to whether biological or environmental factors play a greater role in the male–female differences that have been reported in previous research on throwing?

7. What other differences between boys and girls have been noted in the performance and development of fundamental motor skills? What explanations have been proposed for these differences?

Name_____
Section_____ Date_____

REFERENCES

Hardy, L. L., King, L., Farrell, L., Macniven, R., and Howlett, S. (2010). Fundamental movement skills among Australian preschool children. *Journal of Science and Medicine in Sport, 13*, 503–508.

Haywood, K.M., and Getchell, N. (2009). *Life Span Motor Development* (3rd edition). Champaign, IL: Human Kinetics.

Mitchell, M. (2013). *Introduction to kinesiology: The science of human physical activity.* San Diego, CA: Cognella Academic Publishing. Roberton, M.A., and Halverson, L.E. (1984). *Developing children: Their changing movement.* Philadelphia: Lea and Febiger.

Seefeldt, V., Reuschlein, S., and Vogel, P. (1972). Sequencing motor skills within the physical education curriculum. Paper presented at the annual convention of the American Association for Health, Physical Education, and Recreation, Houston.

Stodden, D. F., Goodway, J. D., Langendorfer, S. J., Roberton, M. A., Rudisill, M. E., Garcia, C., and Garcia, L. E. (2008). A developmental perspective on the role of motor skill competence in physical activity: An emergent relationship. *Quest, 60*, 290–306.

Thomas, J. R., and French, K. E. (1985). Gender differences across age in motor performance: A meta-analysis. *Psychological Bulletin, 98*, 260–282.

Thomas, J. R., and Marzke, M. W. (1992). The development of gender differences in throwing: Is human evolution a factor? In Enhancing Human Performance in Sport: New Concepts and Developments. *Papers from the 63rd Annual Meeting of the American Academy of Physical Education* (pp. 60–76). Champaign, IL: Human Kinetics.

Wild, M. (1938). The behavior pattern of throwing and some observations concerning its course of development in children. *Research Quarterly, 9*, 20–24.

Name_____

Section_____ Date_____

Name_____
Section_____ Date_____

CHAPTER LAB #8.2

TASK ANALYSIS, CLASSIFICATION, AND SKILL

PURPOSE
The aim of this activity is to become familiar with a popular task classification system and to develop skill in classifying tasks and performing task analyses.

INTRODUCTION
Task analysis refers to a specific breakdown of the constraints associated with a particular task followed by a description of the movement pattern that is expected to emerge from those constraints. Task analysis is designed to provide insights into the demands a particular task places on a performer. Once those demands are understood, a practitioner can more easily determine whether an individual has the prerequisites needed to succeed on the task, and how the task might be modified to increase the individual's chances of success. Classification is an important step in the task analysis process because the categories in a classification system provide an initial indication of the generic demands associated with each category. The practitioner can get an immediate sense of how demanding a task is by simply seeing what category it falls into.

In this activity we will use a task analysis procedure designed by Arend and Higgins (1976) to analyze several different tasks. In addition, we will utilize a classification system designed by Gentile (1972, 1987, 2000) that we discussed in Chapter 6 (pages 189–190) of the textbook (Mitchell, 2013) to classify the tasks and to facilitate the task analysis. The classification system examines a task relative to two dimensions: 1) The *environment*, and 2) the *function of the action*. The environment is analyzed relative to what Gentile labels *regulatory conditions* (often called *direct constraints*). Regulatory conditions are the characteristics of objects, surfaces, and other people to which movements must conform to be successful. Tasks are classified according to whether the regulatory conditions are *stationary* or *in motion,* and whether they *vary from trial to trial*. The function of the action is classified as body *stability* or *body transport*. Superimposed on the stability and transport functions is *manipulation*, which, according to Gentile, is action-directed toward maintaining or changing the position of moveable objects.

The table below is an adaptation of the original taxonomy provided by Gentile (1972). It shows the major dimensions of the classification system.

Name_____
Section_____ Date_____

Environmental Context	Regulatory Conditions	Inter-trial Variability	Function of the Action			
			Body Stability		Body Transport	
			No Manipulation	Manipulation	No Manipulation	Manipulation
	Stationary	No ITV				
		ITV				
	In Motion	No ITV				
		ITV				

Below is a sample task analysis for a tennis serve that was originally developed by Susan Higgins (née Arend). We have adapted it for this lab. The steps in the task analysis are as follows:

1. Identify the goal of the task
2. Identify the sub-goals
3. Specify the problems the mover has to solve (link these to the sub-goals)
4. Identify the regulatory conditions
5. Identify whether the regulatory conditions are stationary or in motion
6. Identify the non-regulatory conditions
7. Identify other indirect constraints
8. Identify the function of the action
9. Determine what prerequisites (substrates) the mover needs to be successful

THE TENNIS SERVE

A sample analysis (not meant to be exhaustive)

Goal: To place the ball within the service court in such a way that the opponent is unlikely to return it, using a form that is biomechanically sound and within the limits imposed by the rules of the game.

Sub-Goal(s):

1. Develop a high linear velocity of the racquet head (this sub-goal relates to the function of the action, namely manipulation).

Name_____
Section_____ Date_____

2. Transfer velocity to the ball so that it is well directed and strategically appropriate (this sub-goal relates to the integration of the function of the action—i.e., body stability and manipulation, with the regulatory conditions in the environment).

3. Apply a large force to a moving object while maintaining body stability (this sub-goal relates to the integration of the function of the action—i.e., body stability and manipulation, with the regulatory conditions in the environment).

Note that the goal(s) can be objectively evaluated to determine whether they were achieved. The goal(s) specify the regulatory environmental, biomechanical, and morphological constraints. The goal(s) must be translated to a "force" problem for the performer to solve through movement.

The goals of a tennis serve compel a particular use of the body to summate angular velocity (summate forces) and apply it to the ball—but within the spatial and temporal limits imposed by the ball's flight characteristics and the performer's morphology.

The problems for the mover to solve:

1. Summate forces using a racquet (relates to Sub-Goal 1)

2. Transfer momentum accurately to the ball whose flight characteristics are controlled by the mover (relates to Sub-Goals 2 and 3)

3. Effectively integrate tossing action with racquet action with a dependable consistency (relates to Sub-Goals 2 and 3)

Note that the problems are more abstract than the sub-goals and they cannot be objectively measured. Instead, they describe what the mover must do to achieve the sub-goals.

Regulatory Conditions (Direct Environmental Constraints):
- The distance from the server to the service court
- The height of the net
- The size of the service court
- The height of the ceiling
- The floor surface
- The size, weight, resiliency, shape, and texture of the ball
- The trajectory (speed and direction) of the tossed ball
- The distance of the ball from the server
- Constraints due to the equipment (the racquet): size, weight, length, surface texture and girth of grip, string stiffness

Stationary or In Motion: The regulatory conditions are in motion because at least one of the constraints (the ball) is in motion, even though the server was responsible for putting the ball in motion. Once the ball has left the hand, the server is no longer free to decide when to initiate the swing of the racquet. When a regulatory condition is in motion and the motion varies from trial to trial, we label these tasks as *open* tasks or skills. When the regulatory conditions are stationary and don't vary from trial to trial, we refer to these tasks as *closed* tasks or skills.

Name_____
Section_____ Date_____

Non-Regulatory Conditions (Indirect Environmental Constraints):
The position of the opponent(s), weather conditions, spectators, color of the ball and court. These factors can all influence performance; however, there is no direct mapping between the movement features and each indirect constraint. In comparison, each regulatory condition (direct constraint) maps directly onto a particular feature of the movement (e.g., displacement of limbs, speed of movement, how much force is required, when force should be applied, in what direction force should be applied, etc.).

Other Indirect Contextual Constraints: (effect strategy and performance)
Time of day
Point in the competition; point in the game
Home or away
Who is refereeing
Experience, skill, and mood of the opponent(s)

Function of the Action: Body stability with manipulation

Prerequisites or Substrates of the Tennis Serve:

Emotional:
- Motivation to perform the task
- Free from fears (injury, negative evaluation from peers) that might preclude participation in the task

Cognitive:
- Understanding of the goal of the task
- Knowledge of rules associated with tennis serve
- Knowledge of strategies for successful performance

Perceptual-motor abilities:
- Judge distances and velocity (trajectory)
- Time onset of force application and speed of movement to a moving object
- Kinesthetic awareness of one's height, length of arm with racquet
- Apply force to a moving object with a racquet
- Maintain dynamic equilibrium while applying force
- Dissipate force
- Use body asymmetrically
- Integrate postural control and limb manipulation

Sub-skills:
- Toss a ball with the non-preferred hand to an appropriate height
- Manipulate a racquet as an extension of the arm
- Summate forces—transfer angular velocity from segment farthest from the segment, ultimately applying the force so that each part is brought into action when the preceding segment has reached its peak velocity and least acceleration

Name_____
Section_____ Date_____

Physical abilities (Fitness) needed:
- Minimum hand, wrist, and shoulder/arm strength to manipulate a long lever—a tennis racquet—and to maintain stiffness during the application of force on the ball
- Minimal muscular endurance, and minimal strength and dynamic flexibility to support maximal range of motion of each participating segment during summation of forces

Before learning a tennis serve, the performer should possess the minimal fitness requirements and be able to: (a) summate forces using an overhand pattern, and (b) toss and hit a moving ball without a racquet.

MATERIALS/METHOD

Participants

Every member of the class should participate in this activity. The activity can be done alone or in groups of approximately five students.

Task and Apparatus

The activity simply requires students to classify different tasks and perform a task analysis; however, classification and analysis can best be accomplished by performing the tasks several times.

Procedure

Use Gentile's taxonomy to classify the following fundamental motor skills:
- Hopping
- Skipping
- Standing long jump
- Throwing a ball to a stationary target
- Throwing a ball to a moving target
- Catching a ball
- Striking a stationary ball with an implement
- Striking a moving ball with an implement

Choose one of these tasks and complete a comprehensive task analysis, such as the example tennis serve provided above.

Data Analysis

The tasks can be placed into the taxonomy table on the second page of this activity. The task analysis can be performed on a separate sheet of paper.

RESULTS

Be prepared to share your results with the rest of the class, and to defend your decisions about where to classify each task within the taxonomy.

Name_____

Section_____ Date_____

Name_____
Section_____ Date_____

DISCUSSION QUESTIONS

1. What other systems could be used to classify movements? What are the advantages and disadvantages of these systems relative to Gentile's taxonomy?

2. How do the demands on attention differ when the regulatory conditions are stationary versus in motion?

3. How do the demands on attention vary when the function of the action is body stability compared to body transport? What happens to attention when manipulation is superimposed on stability or transport?

4. The sport of ice hockey has several demanding skills. One of those skills is skating among a crowd of players while controlling a puck with a stick and then passing the puck to a teammate who might be moving at great speed. How would you use Gentile's taxonomy to simplify this task for a novice or a less experienced player?

5. How might you use the taxonomy to develop prerequisite skills that would help ensure success when first exposed to a new skill that places multiple demands on the performer?

6. How does the type of task analysis presented here compare to traditional task analysis and to Burton and Davis's ecological task analysis?

Name_____
Section_____ Date_____

REFERENCES

Arend, S. (1980). Developing the substrates of skillful movement. *Motor Skills: Theory into Practice, 4,* 3–10.

Arend, S. (1980). Developing perceptual skills prior to motor performance. *Motor Skills: Theory into Practice, 4,* 11–17.

Arend, S. and Higgins, J. R. (1976). A strategy for the classification, subjective analysis, and observation of human movement. *Journal of Human Movement Studies, 2,* 36–52.

Burton, A. W., and Davis, W. E. (1996). Ecological task analysis: Utilizing intrinsic measures in research and practice. *Human Movement Science, 15,* 285–314.

Gentile, A. M. (1972). A working model of skill acquisition with application to teaching. *Quest, 17,* 3–23.

Gentile, A. M. (1987). Skill acquisition: Action, movement, and neuromotor processes. In In: J. H. Carr, R. B. Shepherd, J. Gordon, A. M. Gentile, and J. M. Held (Eds.), Movement sciences: Foundation for physical therapy in rehabilitation (pp. 93–154). Rockville, MD: Aspen.

Gentile A. M. (2000). Skill acquisition: Action, movement, and neuromotor processes. In: J. H. Carr, and R. B. Shepherd (Eds.), Movement sciences: Foundation for physical therapy in rehabilitation (2nd Ed., pp 111–187). Gaithersburg, MD: Aspen Publishers.

Mitchell, M. (2013). *Introduction to kinesiology: The science of human physical activity.* San Diego, CA: Cognella Academic Publishing.

Chapter Nine
Sociocultural Foundations

CHAPTER ACTIVITY #9

THE WALKABILITY OF YOUR NEIGHBORHOOD

PURPOSE
The purpose of this activity is to review material presented in Chapter 9 of the textbook (Mitchell, 2013) and to get more familiar with the "walkability" of your neighborhood.

INTRODUCTION
In Chapter 9 we explored many concepts related to sociocultural factors in kinesiology. We examined several sociocultural theories relevant to kinesiology. The concept of rationalization was examined as it influences society and the participation in physical activity. Finally, a number of sociocultural factors that have an effect on the participation in physical activity were discussed. One important factor touched upon was community involvement in promoting physical activity. It was pointed out that increases in crime and traffic can prevent, or certainly discourage, people from walking or riding their bicycles (U.S. Department of Transportation, 1993, 1994). After answering the chapter summary questions, this activity allows you to describe how conducive your neighborhood is for physical activity.

Name_____
Section_____ Date_____

Name_____
Section_____ Date_____

CHAPTER SUMMARY QUESTIONS

1. If you are planning to develop a fitness program for yourself (Chapter 4), name four sociocultural issues you should consider?

   • _____

   • _____

   • _____

   • _____

2. Using each of the theoretical frameworks discussed in this chapter (functionalism, conflict theory, critical theory, and symbolic interactionism), ask one question about why there are either gender or racial differences in physical activity participation.

   Functionalism question: _____

   Conflict theory question:_____

   Critical theory question: _____

   Symbolic interactionism question:_____

3. Provide three additional examples (not discussed in this chapter) of "revenge effects" on the participation in physical activity:

   • _____

   • _____

   • _____

4. Name and briefly describe one example each of how family, peer groups, and school can affect someone's participation in physical activity.

   Family influence: _____

   Peer group influence: _____

   School influence: _____

Name_____

Section_____ Date_____

5. Using Czikszentmihalyi's (1990) concepts of clarity, centering, choice, and commitment, how can a family help a child become more physically active?

Clarity: _____

Centering: _____

Choice: _____

Commitment: _____

6. Provide one example of each type of discrimination in a sport or other physical activity setting.

Position allocation example: _____

Performance differentials: _____

Rewards and authority structure: _____

Name_____
Section_____ Date_____

CHAPTER ACTIVITY

Equipment and Materials Needed: Access to the Internet

The U.S. Department of Health and Human Services (1996) outlined a number of so-called intervention strategies that can increase the level of physical activity in the community. One concept that has recently emerged to describe how conducive a community is to participation in physical activity is called "walkability."

PROCEDURES

Go to the following website and read the information provided: http://en.wikipedia.org/wiki/Walkability

DISCUSSION QUESTIONS

1. What is the definition of walkability?

2. What are some factors affecting walkability?

3. What are some infrastructure factors affecting walkability?

4. What is one of the best ways to measure the walkability of a block, corridor, or neighborhood?

5. What are some benefits of walkability?

6. Describe several ways to make a community more walkable.

Now go to the following website to determine your neighborhood's "walkability score":
http://www.walkscore.com/
Enter the address where you currently reside and obtain a walkability score.

7. What is your current address's walkability score?

Developmental Foundations

Name_____
Section_____ Date_____

On that same page, scroll down and click on *Street Smart Walk Score*.

8. What scores did your address get for the following amenities?

   Groceries:
   Restaurants and bars:
   Shopping:
   Coffee shops:
   Schools:
   Parks:
   Books:
   Entertainment:
   Banking:
   Pedestrian friendliness:

   Now, scroll back up that same page and click on "Your Commute."
   In the box provided, type in the address of "Where Do You Commute?"

9. According to this website, how long is your commute to work or school by:

   Walking?

   Cycling?

   Car?

10. Briefly describe the elevation changes (hills) in your commute.

11. Does the distance and/or elevation changes help determine what mode of transportation you use for your commute?

12. In general, how would describe the walkability of your neighborhood?

13. What are some interventions that your community could implement to improve the walkability of your neighborhood?

Name_____
Section_____ Date_____

REFERENCES

Mitchell, M. (2013). *Introduction to kinesiology: The science of human physical activity.* San Diego, CA: Cognella Academic Publishing.

U.S. Department of Health and Human Services (1996). Physical Activity and Health: A Report of the Surgeon General. Atlanta: U.S. Department of Health and Human Services, Centers for Disease Control and Prevention, National Center for Chronic Disease Prevention and Health Promotion.

U.S. Department of Transportation (1993). Measures to Overcome Impediments to Bicycling and Walking: The National Bicycling and Walking Study, Case Study No. 4. Washington, DC. Department of Transportation, Federal Highway Administration, Publication No. FHWA-PD-93-031.

U.S. Department of Transportation (1994). Final Report: The National Bicycling and Walking Study: Transportation Choices for a Changing America. Washington, DC. Department of Transportation, Federal Highway Administration, Publication No. FHWA-PD-94-023.

http://www.walkscore.com/

http://en.wikipedia.org/wiki/Walkability

Name_____

Section_____ Date_____

Name_____
Section_____ Date_____

CHAPTER LAB #9

GENDER AND RACE IDEOLOGIES IN SPORTS MEDIA

Copyright © 2011 Depositphotos/artjazz.

Copyright © 2011 Depositphotos/Darrin Henry.

PURPOSE

The purpose of this study is to investigate the different ways the media portray gender and racial ideologies in sports.

INTRODUCTION

In Chapter 9 of the textbook (Mitchell, 2013), we discussed several theories of sociology that can be applied to participation in physical activity. Conflict and critical theories emphasize how the "power" manifested in certain individuals can have a strong influence on other people. For sociologists, the ability to do what you want without being stopped by others is known as *power*. Some of the major sources of power and therefore major sources of inequity are money, prestige, information, body size, and strength. If you look closely at kinesiology contexts (i.e., gyms, playgrounds, hiking and biking trails, athletic fields, etc.), you will see that it is not overt, bullying kinds of power that are typically exhibited, but rather subtle, almost hidden forms of power that most influence physical activity involvement. There are no signs, for example, saying that the cardio room or group exercise classes are for women, but in most cases, it is women who populate these spaces. When men enter these spaces, he may be looked down upon or judged by others. He may also be celebrated by others for breaking the gender barriers. Either way, these responses reflect the cultural knowledge that we all share to some degree. This "knowingness" is also a form of power in society because it reinforces the way societal members think things ought to be organized (Hoffman, 2009).

Gender

Gender is different than sex in that gender is not an identity you have, but rather the set of norms or expectations about how we should behave. These set of norms or expectations are linked to societal understandings of sexuality and procreation. We are assigned a gender role based on our sex at birth; however, genders are not natural, biological categories but are socially assigned. In American society, men typically have more power than women, although this relationship changes with race, socioeconomic status, and sexual orientation. In the field of physical activity,

Name_____
Section_____ Date_____

inequalities prevail because of beliefs about the appropriateness of certain forms of physical activity for each gender (Hoffman, 2009).

Research has suggested that boys are more encouraged to engage in vigorous physical activity while girls may even be punished for doing so. Even when allowed to participate, girls' participation in physical activity is often more regulated by their parents (Mitchell, 2013). For example, a young girl might be given permission to play as long as "she gets back in time to set the table" or "if she takes her little brother or sister with her" (Coakley, 2009; Mitchell, 2013). It has also been found that fathers spend considerably more shared physical activity participation time with their sons than their daughters. But even with all these negative factors, women's participation in sports is on the rise (Mitchell, 2013).

Race and ethnicity

Race is a culturally, historically, and socially defined category of social difference typically marked by phenotypical variance among people. Examples of these phenotypical variances are skin color, hair type, facial features, and body shape. It is important to understand that like gender, race is not a fixed identity or biological category but rather defined based on the characteristics we select. Race is a shifting condition of social life that you experience every day. Similar to race is ethnicity, which refers to cultural heritage. An ethnic group is classified as people who share important and distinct cultural traditions, often developed over many generations. Ethnic markers include language, dialect, religion, music, sport activities and games, and manner of dress. Sometimes race and ethnicity overlap. For example, African Americans are defined as a racial group but also distinct cultural traditions. In the United States, members of racial and ethnic minorities have typically held less power than the white majority. Although, there have been strides made in the 20th century, many difficulties remain. Physical activity is also not immune to these problems (Hoffman, 2009).

Both prejudice and discrimination are rampant in sports. This can be seen by the number of minorities in a certain position (position allocation) within a sport. Figures 9.2 and 9.3 from Mitchell (2013) illustrate the percentages of whites, Hispanics, and African Americans in profession football and baseball. There are many hypotheses that attempt to explain differences in position allocation among whites and minorities. One view is that Hispanics and African Americans are recruited for positions that require speed and quickness while whites are recruited for positions that require leadership, decision-making skills, and dependability (Coakley, 2009; Mitchell, 2013).

Media

Sport and media have enjoyed a very symbiotic relationship in U.S. society. On one hand, the enormous popularity of sport is due to the gross coverage of provided by media. On the other, the media are able to generate enormous sales in circulation based upon their extensive treatment of sport. The role of the athlete in media is also important to sales as well as the promotion of the athlete. The mass coverage of athletes may be symbiotic but may also be biased based on race and gender (Wenner, 1989).

Men's sports are 90% of the athletics covered in the media. The images and narratives in this coverage reproduce traditional ideas and beliefs about gender. Women's sports have had an increase in coverage since the mid-1990s mainly in the Olympics, figure skating, tennis and golf tournaments, and some professional and college basketball games (Coakley, 2009). Overall, the coverage of women's sports receives less than 15%, and sports magazines are slow to cover female athletes and women's sports. But when female athletes are portrayed, they tend to include sexualized images of women in the ads accompanying the coverage of men's sports. This pattern of underrepresentation of women's sports in the media exists worldwide (Coakley, 2009).

Name_____
Section_____ Date_____

Much like gender ideologies, racial and ethnic ideologies and stereotypes influence the media as well. In the early part of the 20th century, black athletes were commonly described with references to the "wild" or the "jungle." The color of their skin was also common when describing or labeling the athlete, such as "the black cobra" (Coakley, 2009). Of course, these stereotypes changed during the latter half of the 20th century. The references to "the jungle" and "the wild" were replaced with racialized code words like "the ghetto" and "the streets" (Coakley, 2009). White announcers commonly describe black athletes as having natural abilities, good instincts, unique physical attributes, and tendencies to be undisciplined players. At the same time, white athletes are described by white announcers as being hard-working, intelligent, highly disciplined, and driven by character rather than instincts (Coakley, 2009).

To be accepted into the dominant cultures, Latinos, Asian Americans, Native Americans, and blacks are constantly reminded to "be like whites" in how they think, talk, and act. It is understood by these athletes that to receive favorable media coverage, they should smile in accommodating ways on camera and during interviews just like Michael Jordan and Magic Johnson did for so many years (Coakley, 2009). But the media also admires these athletes for not forgetting where they come from. This creates tension for these athletes. If the players or the coaches try to tell the story, they are quickly accused of playing the race card. But if their stories are not told, then the white privilege in sports will persist without being recognized (Coakley, 2009).

METHODS/MATERIALS

Participant

The participants for this study will be your fellow classmates.

Task and Apparatus

- Note cards (three per person)
- Pictures of athletes (one male and one female per person) from the last 10 years

PROCEDURE

Each student is to bring in two pictures of a sports athlete, one female and one male. The criterion for each picture is the following: from the last 10 years, from a magazine (online or print), and an individual (solo) shot or image. In groups of three, assess the images based on gender and race (Appendix). Write your group's assessment on the note cards and staple to each image. The researchers will then collect all the images for analysis.

RESULTS

Graph classifications 1–3 according to gender and race

Graph the number of women and men in each of the sport types collected

Name_____
Section_____ Date_____

DISCUSSION QUESTIONS

1. Discuss what Title IX is and how it has affected women's participation in sports.

2. Describe the ties between participation in physical activity and power relationships based on gender and race/ethnicity.

3. Describe the ties between leadership in physical activity programs and power relationships based on gender and race/ethnicity.

4. Describe the ties between physical activity expressiveness and power relationships based on gender and race/ethnicity.

5. How has the portrayal of women changed in sport media over the last 10 years? What are some possible reasons for any changes?

6. How has the portrayal of African Americans changed by white announcers since the early 1900s? Find specific examples by announcers or journalists from the past century.

7. Discuss the firing of the Giants' sport talk-radio host in San Francisco in 2005. What were the circumstances for his firing and the response by the public?

REFERENCES

Coakley, J. (2009). *Sports in society: Issues and controversies.* New York: McGraw Hill.
Hoffman, S. (2009). *Introduction to kinesiology: Studying physical activity.* Champaign, IL: Human Kinetics.
Mitchell, M. (2013). *Introduction to kinesiology: The science of human physical activity.* San Diego, CA: Cognella Academic Publishing.
Wenner, L. (1989). *Media, sports, and society.* Newbury Park, CA: Sage Publications, Inc.

Name_____
Section_____ Date_____

APPENDIX

Classifications:

1. In vs. out of uniform

2. On vs. off the court

3. In active vs. passive poses

4. Race: White, African American, Hispanic/Latino, Asian, Native American

5. Male or female

6. Sport type (baseball, basketball, soccer, golf, etc.)

Coding System:

1. In or Out

2. On or Off

3. A or P

4. W, AA H/L, AS, NA

5. M or F

6. Write sport type

Name_____

Section_____ Date_____

Chapter Ten
Epilogue

CHAPTER ACTIVITY #10

EXPLORING THE AMERICAN KINESIOLOGY ASSOCIATION

Source: American Kinesiology Association.

PURPOSE
The purpose of this activity is to review material presented in Chapter 10 of the textbook (Mitchell, 2013) and to investigate some of the important positions taken by the primary national organization representing the field of kinesiology, the American Kinesiology Association (AKA).

INTRODUCTION
In Chapter 10, we explored some ideas related to how research and teaching in kinesiology can adopt a more integrative approach that utilizes information, concepts, and principles within the various subfields of kinesiology. The argument was made that the unique contribution of kinesiology occurs when information from more than one subfield can be used to enhance teaching and research. In addition, we described the two major approaches to advancing knowledge within kinesiology: positivism and holism, and pointed out the unique contribution that phenomenology can give to understanding the importance of physical activity. We also briefly discussed some issues related to the future of kinesiology. After answering some chapter summary questions, this activity will allow you to learn more about the American Kinesiology Association and its role in representing kinesiology.

Name_____

Section_____ Date_____

CHAPTER SUMMARY QUESTIONS

1. Provide an example of the positivism approach to the study of exercise physiology, motor control/learning, and sport and exercise physiology.

    Exercise physiology: _____

    Motor control: _____

    Sport and exercise physiology: _____

2. Provide three reasons that an integrative approach to kinesiology is useful:

    - _____

    - _____

    - _____

3. Give two examples of integrative teaching in kinesiology:

    - _____

    - _____

4. Describe one major difference between the reductionist and holistic approach to the study of kinesiology:

    _____

    _____

5. How is the phenomenological approach different from the empirical approach as it applies to the study of kinesiology?

    _____

    _____

    _____

Name_____
Section_____ Date_____

CHAPTER ACTIVITY

Equipment and Materials Needed: Access to the Internet

This activity will require you to go to the website of the AKA and read about some important positions taken by the organization. After reading each position statement you will be required to answer a few questions. This activity will hopefully allow you to get a better understanding of the role the AKA plays in representing kinesiology.

Go the main website of the AKA:

http://www.americankinesiology.org

Click on "Publications" and then click on "White Papers"

Read the following white papers:
1. AKA Clarifies the Definition of Kinesiology
2. Kinesiology on the Move: One of the Fastest Growing (But Often Misunderstood) Majors in Academia
3. The American Kinesiology Association and the Future of Kinesiology
4. Re-Examining the Undergraduate Core in Kinesiology in a Time of Change

DISCUSSION QUESTIONS

1. Which papers advocated that the field of kinesiology should be integrative or cross-disciplinary? In what way?

2. If you had to explain to someone unfamiliar with the field, how would you define kinesiology?

3. How popular with students is the study of kinesiology?

4. Describe at least three major points made by Dr. Hal Lawson about the future of kinesiology.

5. What are the major responsibilities of the AKA?

6. According to the AKA, what are four major elements of the undergraduate kinesiology core?

REFERENCES

Mitchell, M. (2013). *Introduction to kinesiology: The science of human physical activity.* San Diego, CA: Cognella Academic Publishing.

http://www.americankinesiology.org

Name_____

Section_____ Date_____

Name_____
Section_____ Date_____

CHAPTER LAB #10

DESIGNING AN INTEGRATIVE RESEARCH PROJECT

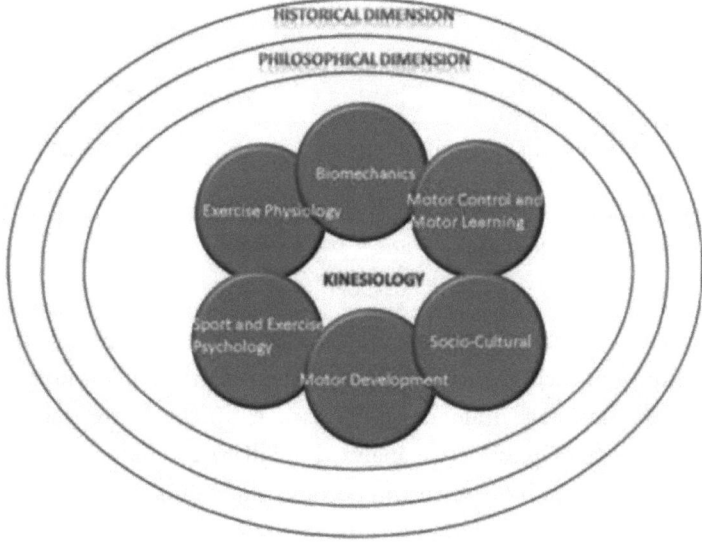

PURPOSE

The aim of this activity is to design an integrative research project. This activity will challenge you to incorporate some of the concepts, theories, and measurement methods introduced in this textbook from the various subfields of kinesiology.

INTRODUCTION

Mitchell (2013) has noted that physical activity is "…the result of complex interactions of sociocultural, psychological, physiological, biomechanical, and anatomical factors" (p. 302). In other words, although we often examine physical activity from the perspective of only one of these subfields of kinesiology, we are also aware that certain concepts from many of the subfields will influence any given physical activity in terms of cognitions, performance, learning, and/or participation. Thus, and similar to the point raised in your Chapter 1 lab, it is beneficial for us to examine a kinesiological phenomenon from many different angles (i.e., from various subfields) *within the same study or experiment*. If in a kinesiological examination we recognize the influence of, and systematically incorporate, factors from multiple subfields, we then are more likely to strike a reasonable balance between two of the major goals of any scientific inquiry: achieving a reasonable degree of experimental control, and of ecological validity (Schmuckler, 2001). If this reasonable balance is accomplished, we will obtain a more comprehensive, more complex, and more accurate understanding of human physical activity.

Name_____
Section_____ Date_____

MATERIALS/METHOD

Participants

Students will work in groups of two or three to complete this lab assignment. Every member of the class should participate.

Task and Apparatus

The activity requires students to design an integrative research project that examines a specific kinesiology-related research question.

Procedure

Recall your Chapter 1 lab where you were required to find three peer-reviewed journal articles (each from a different subfield), summarize the findings of each journal article, and then clearly explain how those findings contributed to our understanding of factors that may influence cardiovascular fitness. In this lab, you will be constructing a research project that examines the same question we analyzed in Chapter 1: "How does one improve cardiovascular fitness?" In this lab, however, your research question will become much more refined, and instead of reporting what we *already know* about this topic, you will be designing a research project to *add to what we know*!

When researchers conceptualize a research project, an important first step after the general topic of interest has been identified (e.g., "improving cardiovascular fitness") is to conduct a literature review of the current body of research literature related to that topic (similar to your Chapter 1 lab, just much more extensive). This serves two functions: it identifies what we already know about a topic, *and* it identifies what we do *not* know (i.e., gaps, limitations, or unanswered questions about the topic). It is the things that we do *not* know that guide us in creating a more refined research question. The goal of most researchers is to complete a research project that provides us with novel information—essentially, conclusions or relationships that have not been previously established. This is how we are able to continue to expand our understanding of a real-life phenomenon, such as physical activity.

Data Analysis

Retrieve <u>TWO</u> of the three journal articles you used for your Chapter 1 lab (your choice of which two articles). Use the "Data Analysis" table below to help you organize the relevant information from your two articles. We recommend you reproduce this table in a spreadsheet program (e.g., Microsoft Excel) so that you can add rows as needed.

For the data analysis portion of this lab, you are required to review the introduction section of your two journal articles in great detail. You are looking for the authors' summaries of things that we *already know* about improving cardiovascular fitness—things that previous research have already established. Record each relevant piece of information (i.e., information about improving cardiovascular fitness) you find in the Data Analysis table below. Remember to also record what each current study or experiment found in terms of improving cardiovascular fitness (their findings are also something that we "already know" about our topic).

Name_____

Section_____ Date_____

Data Analysis Table

Article you are getting this information from (cite):	What we know about improving cardiovascular fitness:	Subdiscipline(s) of kinesiology this relates to:

Note: Add more rows as needed ... you *should* find that you need to add more rows!

RESULTS

For this lab, you are required to do a bit more thinking to come up with your "results." You are not reporting numbers here! Instead, you and your groupmates will critically analyze your summary of what we already know about improving cardiovascular fitness (based on the information you collected from your two journal articles) and identify what we *do not* know, as it relates to the specific subfields and the specific cardiovascular fitness-related factors reviewed in your summary.

You are required to come up with two "unknowns" about improving cardiovascular fitness—one unknown for each subfield you examined (recall that each journal article had to come from a different subfield of kinesiology). Note that the unknowns you identify *must* be derived from the specific information in your summary of the current literature. It may be helpful to think of this task of "identifying the unknowns" as you being challenged to find the *limitations, gaps, or unanswered questions* in our understanding of improving cardiovascular fitness, *specific* to the information you have gathered for each of your two journal articles. Frame each unknown as a question (e.g., Question 1: "What is the influence of social physique anxiety on regular participation in group exercise classes?"; Question 2: "Does *frequency* of exercise, rather than intensity and duration, affect cardiovascular fitness?"). List your two research questions (i.e., the unknowns) below the Data Analysis Table you create.

DISCUSSION QUESTIONS

1. Explain your rationale for coming up with the research questions that you did in your results section (i.e., explain how you decided on those two questions).

2. Attempt to combine your two separate research questions you developed in your results section into one single research question. For example, one could possibly combine Questions 1 and 2 in our example from the results section to create this single research question: "Does social physique anxiety affect group exercise class attendance and cardiovascular fitness levels?"

Name_____
Section_____ Date_____

3. Loosely outline how you might design an experiment to test the research question you constructed in discussion Question 2. Be sure to include who your participants will be, what task they will be doing, what your different experimental groups are, and how you will measure each dependent variable (i.e., the variables that you will use to compare the groups to each other).

4. Based on the information you gleaned through this activity, why (in your own words) is it important to sometimes take a broader perspective when trying to learn about a particular topic or issue?

5. As it relates to what we understand about cardiovascular fitness, what do you think is a potential limitation of only reading one article (versus many) from each subfield?

6. Compare your responses to discussion Questions 4 and 5 in this lab to your responses to discussion Questions 1 and 4 in your Chapter 1 lab. Were there any notable differences between how you responded to these questions then versus now? What do you think caused these differences? What can you learn from this reflection and comparison task?

REFERENCES

Mitchell, M. (2013). *Introduction to kinesiology: The science of human physical activity.* San Diego, CA: Cognella Academic Publishing.

Schmuckler, M. A. (2001). What is ecological validity? A dimensional analysis. *Infancy, 2,* 419–436.

Printed by Libri Plureos GmbH in Hamburg, Germany